# THE FIELD DESCRIPTION OF SEDIMENTARY ROCKS

W9-DEW-390

The Geological Society of London Handbook Series
published in association with
The Open University Press comprises

Barnes: *Basic Geological Mapping*
Tucker: *The Field Description of Sedimentary Rocks*
Fry: *The Field Description of Metamorphic Rocks*
  (in preparation)
Thorpe and Brown: *The Field Description of Igneous Rocks*
  (in preparation)

Geological Society of London Handbook
HANDBOOK SERIES EDITOR–M.H. de FREITAS

# The Field Description of

# Sedimentary Rocks

### Maurice E. Tucker

*Department of Geology*
*University of Newcastle-upon-Tyne*

THE OPEN UNIVERSITY PRESS

MILTON KEYNES

and

HALSTED PRESS

John Wiley & Sons

New York – Toronto

First published 1982 by
The Open University Press
A Division of
Open University Educational Enterprises Limited
12 Cofferidge Close
Stony Stratford
Milton Keynes, MK11 1BY
England

Printed in Great Britain by
Redwood Burn Ltd, Trowbridge

British Library Cataloguing in Publication Data

Tucker, Maurice E.
    The field description of sedimentary rocks.
    –(Geological Society of London handbook series; 2)
    1. Rocks, Sedimentary 2. Geology – Field Work
    I.    Title II. Series
    552'.5    QE604
    **ISBN 0 335 10036 8 (Open University Press)**

Published in the U.S.A., Canada and Latin America by
Halsted Press, a Division of John Wiley & Sons, Inc.,
New York.

Library of Congress Cataloging in Publication Data

Tucker, Maurice E.
    The Field Description of Sedimentary Rocks

    (Geological Society of London handbook series)
    "A Halsted Press Book."
    Bibliography: p. 111
    1. Rocks, Sedimentary.    2. Petrology–Fieldwork.
    I. Title.    II. Series.
    QE471.J83        552'.5        81-6539
    ISBN 0-470-27239-2 (Halsted)    AACR2

# Contents

# *Preface*

The study of sedimentary rocks is often an exciting, challenging and rewarding occupation. However, to get the most out of these rocks, it is necessary to undertake precise and accurate fieldwork. The secret of successful fieldwork is a keen eye for detail and an enquiring mind. Be observant, see everything in the outcrop, then think about the things seen and look again. This book is intended to show how sedimentary rocks are tackled in the field and has been written for those with a geological background of at least first year university or equivalent.

At the outset, this book describes how the features of sedimentary rocks can be recorded in the field, particularly through the construction of graphic logs. The latter technique is widely used since it provides a means of recording all details in a handy form; further, from the data, trends through a sequence and differences between horizons readily become apparent. In succeeding chapters, the various sedimentary rock types, textures and structures are discussed as they can be described and measured in the field. A short chapter deals with fossils since these are an important component of sedimentary rocks and much useful information can be derived from them for environmental analysis; they are also important in stratigraphic correlation and palaeontological studies. Having collected the field information there is the problem of knowing what to do with it. A concluding section deals briefly with facies identification and points the way towards facies interpretations.

Maurice E. Tucker

# Acknowledgements

I should like to thank the many friends and colleagues who have willingly read drafts of this handbook and kindly provided photographs. I am indebted to Vivienne for typing the manuscript and helping in the other ways which only a wife can.

# 1
# Introduction

This book aims to provide a guide to sedimentary rocks in the field. It describes how to recognize the common lithologies, textures and sedimentary structures, and how to record and measure these features.

## 1.1 Tools of the trade

Apart from a notebook (size around 10 × 20 cm), pens, pencils, appropriate clothing, footwear and a rucksack, the basic equipment of a field geologist comprises a hammer, chisel, handlens, compass-clinometer, tape measure or steel rule, acid-bottle, sample bags and felt-tip pen. A camera is invaluable. Topographic and geological maps should also be carried, as well as any pertinent literature.

For most sedimentary rocks, a geological hammer of around 1 kg (2 lbs) is sufficiently heavy. A range of chisels can be useful, if a lot of collecting is anticipated. A handlens is an essential piece of equipment; × 10 magnification is recommended since with this grains and features down to 100 microns and less can be observed. To become familiar with the size of grains as seen through a handlens, examine the grains against a ruler graduated in mm or half mm. A compass-clinometer is also important

for taking routine dip and strike and other structural measurements, and also for measuring palaeocurrent directions: correct the compass for the angle between magnetic north and true north. This angle of declination is normally given on topographic maps of the region. You should also be aware that power lines, pylons, metal objects (such as your hammer) and some rocks (although generally mafic-ultramafic igneous bodies) can affect the compass reading, and produce spurious results. A tape or steel rule, preferably several metres in length, is necessary for measuring the thickness of beds and dimensions of sedimentary structures. For the identification of calcareous sediments a plastic bottle of hydrochloric acid (around 10–20%) is useful, and if some alizarin Red S is added, then dolomites can be distinguished from limestones. Polythene or cloth bags for samples and a felt-tip pen (preferably with waterproof, quick drying ink) for writing numbers on the specimens are also necessary. Friable specimens and fossils should be carefully wrapped in newspaper to prevent breakage.

If unconsolidated rocks or modern sediments are being studied you need a trowel and spade. Epoxy-resin-cloth peels can be made in the field of vertical sections through soft sediments. The techniques for taking such peels are given in Bouma (1969).

Other non-geological items which it is useful to carry in the rucksack include: a whistle, first aid equipment, matches, emergency rations, a knife, waterproof clothing and a 'space blanket'.

# 2

# *Field techniques*

## 2.1 What to look for

There are six aspects of sedimentary rocks which it is necessary to consider in the field, and which should be recorded in as much detail as possible. These are: the *lithology*, that is the composition and/or mineralogy of the sediment; the *texture*, referring to the features and arrangements of the grains in the sediment, of which the most important aspect to examine in the field is the grain-size; the *sedi-mentary structures*, present on bedding surfaces and within beds, some of which record the *palaeo-currents* which deposited the rock; the *colour* of the sedimentary rock; the *thickness* and *geometry* of the beds or rock units and of the sedimentary rock mass as a whole; and the nature, distribution and preservation of *fossils* contained within the sedimentary rocks. A broad scheme for the study of sedimentary rocks in the field is given in Table 2.1.

**Table 2.1** Broad scheme for the study of sedimentary rocks in the field, together with reference to appropriate chapters in this book

A   Identify lithology by establishing mineralogy/composition of rock; see Chapter 3.

B   Examine texture of rock: grain shape and roundness, sorting, fabric and colour; see Chapter 4.

C   Look for sedimentary structures on bedding surfaces and undersurfaces, and within beds; see Chapter 5.

D   Deduce the geometry of the sedimentary rock beds, units and bodies; see Section 5.7.

E   Search for fossils and note types present, modes of occurrence and preservation; see Chapter 6.

F   Measure all structures giving palaeocurrent direction; see Chapter 7.

G   Record details of sequence by means of graphic log and notes and sketches in field notebook; see Chapter 2.

H   Consider, perhaps at later date, lithofacies present, depositional processes, environmental interpretations and palaeogeography; see Chapter 8.

I   Undertake laboratory work to confirm and extend field observations on rock composition/mineralogy, texture, structures, fossils (etc.) and to pursue other lines of enquiry such as on the diagenesis and geochemistry of the sediments.

The various attributes of a sedimentary rock combine to define a *facies*, which is the product of a particular depositional environment or depositional process in that environment. Facies identification and facies analysis are the next steps after the field data have been collected. These topics are briefly discussed in Chapter 8.

## 2.2 The approach

The question of how many exposures to examine per square kilometre depends on the aims of the study, the time available, the lateral and vertical facies variation and the structural complexity of the area. If it is a reconnaissance survey of a particular formation or group then well-spaced sections will be necessary; if a specific member or horizon is being studied then all available outcrops will need to be looked at; individual beds may have to be followed laterally.

The best approach at outcrops is initially to survey the rocks from a distance, noting the general relationships and any folds or faults which are present. Some larger-scale structures, such as channels and erosion surfaces, and the geometry of sedimentary rock units, are best observed from a distance. Then take a closer look and see what lithologies and lithofacies are exposed. Check the way-up of the strata using sedimentary structures such as cross-bedding, graded bedding, scours, sole structures, geopetals in limestones, or cleavage/bedding relationships.

Having established approximately what the outcrop has to offer, decide whether the section is worth describing in detail. If it is, it is best to record the sequence in the form of a graphic log (Section 2.4). If the exposure is not good enough for a log, then notes and sketches in the field notebook will have to suffice. In any event, not all the field information can go on the log.

## 2.3 Field notes

Your notebook should be kept as neat and well-organized as possible. The location of sections being examined should be given precisely, with a grid reference and sketch map. Any relevant stratigraphic information should also be entered. It is easy to forget such things with the passage of time. Incidental facts jotted down, such as the weather, or a bird seen, can jolt the memory about the locality in years to come when looking back through the book.

Notes written in the field book should be factual, accurately describing what you are looking at, without any interpretations; these can come later when the field data are analyzed. Describe the size, shape, orientation of the features. Make neat and accurate labelled sketches, with a scale. When taking photographs do not forget to put in a scale. Record the location and subject of photographs in the notebook.

One attribute of the sediments which cannot be recorded adequately on the log is the geometry of the bed or rock unit and of the sedimentary rock mass as a whole (Section 5.7). Sketches, photographs and descriptions should be made of the shape and lateral changes in thickness of beds as seen in quarry and cliff faces. Rock units may also change facies laterally. Local detailed mapping and logging of many small sections may be required in areas of poor exposure to deduce lateral changes.

## 2.4 Graphic logs

The standard method for collecting field data of sedimentary rocks is to construct a graphic log of the sequence (Figs. 2.1 and 2.2). They immediately give a visual impression of the section, and are a convenient way of making correlations and comparisons between equivalent sections from different areas; repetitions, cycles and general trends may become apparent.

The vertical scale used depends on the detail required and available. For precise work, 1:10 or 1:5 is used but for many purposes 1:50 (that is 1 cm on the log equals 0.5 metre) or 1:100 (1 cm equals 1 metre) is adequate.

There is no set format for a graphic log; indeed, the features which can be recorded do vary from sequence to sequence. Features which it is necessary to record and which therefore require a column on the log are: bed or rock unit thickness; lithology; texture, especially grain-size; sedimentary structures; palaeocurrents; colour; and fossils. The nature of bed contacts can also be marked on the log. A further column for special or additional features ('remarks') can also be useful. If you are going to spend some time in the field then it is worth preparing the log sheets before you go. An alternative is to construct a log in your field notebook, but this is less satisfactory since the page size of most notebooks is too small.

**Fig. 2.1** An example of a graphic log; symbols are given in Fig. 2.2.

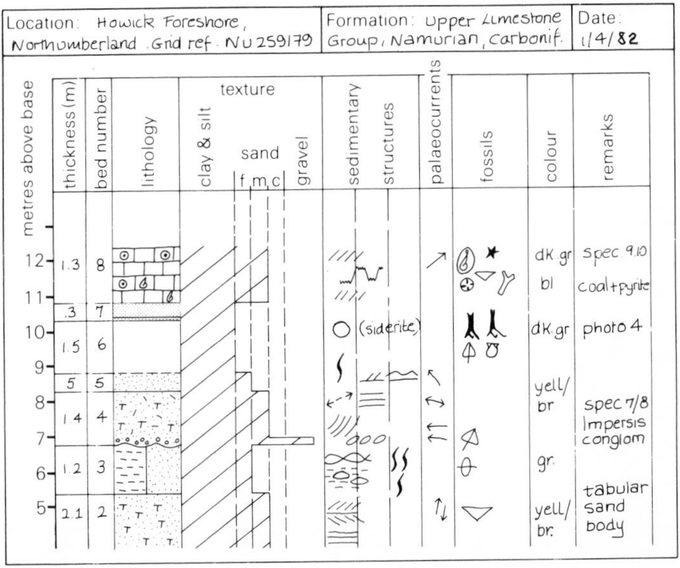

| Location: Howick Foreshore, Northumberland .Grid ref. NU 259179 | | | | | | | Formation: Upper Limestone Group, Namurian, Carbonif. | | | | Date: 1/4/82 |

| metres above base | thickness (m) | bed number | lithology | texture clay & silt | sand f m c | gravel | sedimentary structures | palaeocurrents | fossils | colour | remarks |
|---|---|---|---|---|---|---|---|---|---|---|---|
| 12 | 1.3 | 8 | | | | | | ↗ | ⊘ ✱ | dk gr | spec 9,10 |
| 11 | | | | | | | | | ⊕ ▽ γ | bl | coal+pyrite |
| | .3 | 7 | | | | | | | | | |
| 10 | 1.5 | 6 | | | | | ○ (siderite) | | ↕ ↕ △ ʊ | dk gr | photo 4 |
| 9 | | | | | | | | | | | |
| 8 | 5 | 5 | | | | | | ↖ | | yell/br | spec 7/8 |
| | 1.4 | 4 | | | | | ←·→ | ↩ | ↔ | | impersis |
| 7 | | | | | | | 000 | ⟵ | ⌀ | | conglom |
| 6 | 1.2 | 3 | | | | | | ⟨⟨ | ⊖ | gr | |
| 5 | 2.1 | 2 | | | | | | ↿↓ | ▽ | yell/br | tabular sand body |

13

Where exposure is continuous or nearly so, then there is no problem concerning the line of the log; simply take the easiest path. If outcrop is good but not everywhere continuous it may be necessary to move laterally along the section to find outcrops of the succeeding beds. Some small

**Fig. 2.2** Symbols for lithology, sedimentary structures and fossils for use in a graphic log.

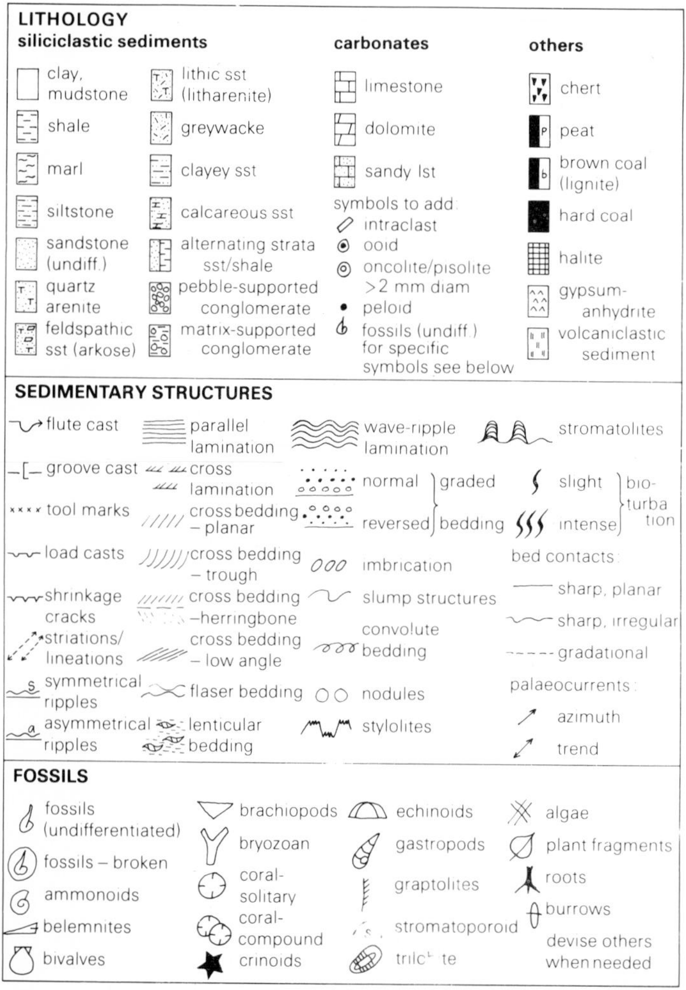

## LITHOLOGY

### siliciclastic sediments

| | | | |
|---|---|---|---|
| clay, mudstone | lithic sst (litharenite) | | |
| shale | greywacke | | |
| marl | clayey sst | | |
| siltstone | calcareous sst | | |
| sandstone (undiff.) | alternating strata sst/shale | | |
| quartz arenite | pebble-supported conglomerate | | |
| feldspathic sst (arkose) | matrix-supported conglomerate | | |

### carbonates

limestone

dolomite

sandy lst

symbols to add:
⌀ intraclast
◉ ooid
◎ oncolite/pisolite >2 mm diam
• peloid
б fossils (undiff.) for specific symbols see below

### others

chert

peat

brown coal (lignite)

hard coal

halite

gypsum-anhydrite

volcaniclastic sediment

## SEDIMENTARY STRUCTURES

flute cast

groove cast

tool marks

load casts

shrinkage cracks

striations/ lineations

symmetrical ripples

asymmetrical ripples

parallel lamination

cross lamination

cross bedding – planar

cross bedding – trough

cross bedding – herringbone

cross bedding – low angle

flaser bedding

lenticular bedding

wave-ripple lamination

normal ⎫ graded
reversed ⎭ bedding

imbrication

slump structures

convolute bedding

nodules

stylolites

stromatolites

slight ⎫ bio-
intense ⎭ turbation

bed contacts:
— sharp, planar
— sharp, irregular
- - - - gradational

palaeocurrents:
azimuth
trend

## FOSSILS

| | | | |
|---|---|---|---|
| fossils (undifferentiated) | brachiopods | echinoids | algae |
| fossils – broken | bryozoan | gastropods | plant fragments |
| ammonoids | coral-solitary | graptolites | roots |
| belemnites | coral-compound | stromatoporoid | burrows |
| bivalves | crinoids | trilcᵗte | devise others when needed |

14

excavations may be required where rocks in the sequence, often mud-rocks, are not exposed; otherwise enter 'no exposure' on the log. It is best to log from the base of the sequence upwards.

## 2.4.1 Bed/rock unit thicknesses

These are measured with a tape measure; care must be exercised where rocks dip at a high angle and the exposure surface is oblique to the bedding. Attention needs to be given to where boundaries are drawn between units in the sequence; if there are obvious bedding planes or changes in lithology then there is no problem. Thin beds, all appearing identical, can be grouped together in a single lithological unit, if the log has a small scale. Where there is a rapid alternation of thin beds of differing lithology, they can be treated as one unit and notes made of the thicknesses and character of individual beds noting any increases or decreases in bed thickness up the sequence. It is often useful to give each bed or rock unit a number so as to facilitate later reference beginning at the strati-graphically lowest bed.

## 2.4.2 Lithology

On the graphic log, this is recorded in a column by using an appropriate ornamentation, Fig. 2.2. If it is possible to subdivide the lithologies further, then more symbols can be added, or coloured pencils used. If two lithologies are thinly interbed-ded, then the column can be divided into two by a vertical line and the two types of ornament entered. More detailed comments and observations

on the lithology should be entered in the field notebook, reference to the bed or rock unit being made by its number.

## 2.4.3 Texture (grain-size)

On the log there is a horizontal scale (the textural column), showing clay and silt, sand (divided into fine, medium and coarse) and gravel. Gravel can be divided further if coarse sediments are being logged. To aid the recording of grain-size (or crystal-size), fine vertical lines can be drawn for each grain-size class boundary. Having determined the grain-size of a rock unit, mark this on the log and shade the area; the wider the column, the coarser the rock. Orna-ment for the lithology and/or sedi-mentary structures can be added to this textural column. Other textural features, such as grain fabric, round-ness and shape, should be recorded in the field notebook, although distinc-tive points can be noted in the remarks column. Particular attention should be given to these features if conglomerates and breccias are in the sequence (Section 4.6).

## 2.4.4 Sedimentary structures and bed contacts

Sedimentary structures and bed con-tacts present in the rock sequence can be recorded in a column by symbols. Sedimentary structures occur on the upper and lower surfaces of beds as well as within them. Thus separate columns can be drawn up for surface and internal sedimentary structures if they are both common. Symbols for the common sedimentary structures are shown in Fig. 2.2. Measurements, sketches and descriptions of the structures should be made in the field notebook.

Note whether boundaries are (a) sharp and planar, (b) sharp and scoured or (c) gradational: each can be represented in the lithology column by a straight, irregular or dashed line respectively.

### 2.4.5 Palaeocurrent directions

For the graphic log, these can be entered either in a separate column or adjacent to the textural log as an arrow or trend line. The measurements themselves should be retained in the field notebook.

### 2.4.6 Fossils

Fossils indicated on the graphic log record the principal fossil groups present in the rocks. Symbols which are commonly used are shown in Fig. 2.2. These can be placed in a fossil column alongside the sedimentary structures. If fossils make up much of the rock (as in some limestones) then the symbol(s) of the main group(s) can be used in the lithology column. Observations on the fossils themselves should be entered in the field notebook (Chapter 6).

### 2.4.7 Colour

The colour of a sedimentary rock is best recorded by use of a colour chart, but if this is not available then simply devise abbreviations for the colour column.

### 2.4.8 'Remarks' column

This can be used for special features of the bed or rock unit, such as degree of weathering and presence of authigenic minerals (pyrite, glauconite, etc.) and supplementary data on the sedimentary structures, texture or lithology. Specimen numbers can be entered here and the location of photographs or of sketches in your notebook.

## 2.5 Presentation of results

Once the field data have been collected it is useful to consider how these may be presented or communicated to others. Two common schemes involve summary graphic logs and lithofacies maps.

A summary log generally consists of one column depicting the grainsize, principal sedimentary structures and broad lithology; this gives an immediate impression of the nature of the rock sequence, Fig. 2.3. If it is necessary to give more information then lithology can be represented in an adjacent column alongside the summary log.

A lithofacies map shows the distribution of lithofacies of laterally-equivalent strata over an area. Maps can be drawn to show variations in specific features of the facies, such as sediment grain-size, thickness and sandstone/shale ratio.

## 2.6 Collecting specimens

For much sedimentological laboratory work, samples of hand specimen-size are sufficient, although this does depend on the nature of the rock and on the purpose for which it is required. Samples should be of in situ rock and you should check that they are fresh, unweathered, and represen-

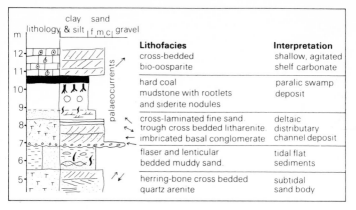

| | Lithofacies | Interpretation |
|---|---|---|
| | cross-bedded bio-oosparite | shallow, agitated shelf carbonate |
| | hard coal mudstone with rootlets and siderite nodules | paralic swamp deposit |
| | cross-laminated fine sand. trough cross bedded litharenite. imbricated basal conglomerate | deltaic distributary channel deposit |
| | flaser and lenticular bedded muddy sand. | tidal flat sediments |
| | herring-bone cross bedded quartz arenite | subtidal sand body |

**Fig. 2.3** An example of a summary graphic log, based on data of Fig. 2.1.

tative of the lithology. Do not forget to label the rock sample; give it (and its bag) a number using a waterproof felt-tip pen. In many cases, it is useful or necessary to mark the way-up of the specimen; an arrow pointing to the stratigraphic top is sufficient for this. For detailed fabric studies, the orientation of the sample (strike and dip) should also be marked on the sample. As a safeguard, specimen number and orientation data can be recorded in the field notebook, with a sketch of the specimen.

Specimens can also be collected for extraction of microfossils—such as foraminifera from Mesozoic-Cainozoic mudrocks and conodonts from Palaeozoic limestones. A hand-sized sample is usually sufficient for a pilot study. Macro-fossils too can be collected in the field, for later cleaning up and identification. It is best not to collect just for the sake of it: only take away what is really necessary for your project. Ensure you collect representatives of the whole fauna. Faunas from different beds or lithofacies should be kept in separate bags. Many fossils will need to be individually wrapped in newspaper.

## 2.7 Stratigraphic practice

Stratigraphically, rocks are divided up on the basis of lithology (lithostratigraphy), fossils (biostratigraphy) and time (chronostratigraphy). From field studies, sedimentary rocks are primarily considered in purely descriptive lithostratigraphic terms. The fundamental unit in lithostratigraphy is the *formation*, possessing an internal lithological homogeneity and serving as a basic mappable unit. Adjacent formations should be readily distinguishable on physical or palaeontological characters. Boundaries may be gradational, but they should be clearly, even if arbitrarily, defined in a designated type section or sections. Although thickness is not a criterion, formations are typically $10^2$–$10^3$ m thick. Thickness will vary laterally over an area and formations are often diachronous on a large scale. Stratigraphically adjacent and related formations, such as deposited within the same basin, may be associated so as to constitute a *group* (typically $10^3$ m+ thick). A formation may be wholly or partially subdivided into *members*, characterized by more particular lithological features, and if there is a distinctive *bed* within a member this too can be named. Lithostratigraphic units are given geographical names with capital initials.

To erect a lithostratigraphy yourself the International Stratigraphic Guide, edited by Hedberg (1976) or the Guide to Stratigraphical Procedure by Holland *et al.* (1976) should be followed: in many parts of the world, older stratigraphic names are in use which do not conform with the International Code.

# Sedimentary rock types

## 3.1 Principal lithological groups

For the identification of sedimentary rock types in the field the two principal features to note are composition-mineralogy and grain-size.

On the basis of origin, sedimentary rocks can be grouped broadly into four categories (Table 3.1).

The most common lithologies are the sandstones, mudrocks and limestones (which may be altered to dolomites). Other types are often only locally well developed.

In some cases you may have to think twice as to whether the rock is sedimentary in origin or not. Points generally indicating a sedimentary origin include:

1. the presence of stratification;
2. the presence of sedimentary structures on bedding surfaces and within beds;
3. the presence of fossils;
4. the presence of grains which have been transported (that is, clasts);
5. the presence of minerals invariably of

sedimentary origin (e.g., glauconite, chamosite).

### 3.1.1 Terrigenous clastic rocks

These are dominated by detrital grains (silicate minerals and rock fragments especially) and include the *sandstones, mudrocks, conglomerates* and *breccias*. Sandstones are composed of grains chiefly between 1/16 and 2 mm in diameter (Section 3.2). Bedding is usually distinct and internal, and bedding plane, sedimentary structures are common.

Conglomerates and breccias (Section 3.3) consist of large clasts (pebbles, cobbles and boulders), more rounded in conglomerates, more angular in breccias, with or without a sandy or muddy matrix.

Mudrocks (Section 3.4) are fine grained, particles mostly less than 1/16 mm diameter, and dominated by clay minerals and silt-grade quartz. Many mudrocks are poorly bedded,

**Table 3.1** The four principal categories of sedimentary rock (1–4) together with the broad lithological groups

| 1 terrigenous clastics | 2 biochemical-biogenic -organic deposits | 3 chemical precipitates | 4 volcaniclastics |
|---|---|---|---|
| sandstones, conglomerates and breccias, mudrocks | limestones + dolomites, chert, phosphates, coal | ironstones, evaporites | tuffs, agglomerates |

and also poorly exposed. Colour is highly variable, as is fossil content.

### 3.1.2 Limestones (Section 3.5)

Limestones are composed of more than 50% $CaCO_3$ and so the standard test is to apply dilute acid (HCl); the rock will fizz. Many limestones are a shade of grey but all other colours occur. Fossils are fequently present, often in profusion. Dolomites or dolostones are composed of more than 50% $CaMg(CO_3)_2$. They react little with dilute acid but more readily with hot or more concentrated acid. Alizarin red S in HCl stains limestone pink to mauve, whereas dolomite is unstained. Many dolomites are creamy yellow or brown in colour and they are often harder than limestones. Most dolomites have formed by replacement of limestones through which original structures have been partly or wholly destroyed. Poor preservation of fossils may thus indicate dolomitization.

### 3.1.3 Other lithologies

The only evaporite deposit (Section 3.6) occurring commonly at surface outcrop is gypsum, mostly as nodules of fine crystals, although veins of fibrous gypsum (satin spar) are usually associated. Evaporites such as anhydrite and halite are only encountered at the surface in very arid areas.

Ironstones (Section 3.7) include bedded, nodular, oolitic and replacement forms. They frequently weather to a rusty yellow or brown colour in surface outcrops. Some ironstones feel heavy relative to other sediments.

Cherts (Section 3.8) are mostly cryptocrystalline to microcrystalline siliceous rocks, occurring as very hard, bedded units or nodules in other lithologies (particularly limestones): many cherts are dark grey to black, or red.

Sedimentary phosphate deposits (Section 3.9) (phosphorites) frequently consist of concentrations of bone fragments and/or phosphate nodules. The phosphate itself is usually cryptocrystalline, dull on a fresh fracture surface with a brownish or black colour.

Organic sediments (Section 3.10) such as hard coal, brown coal (lignite) and peat should be familiar and oil shale can be recognized by its smell and dark colour.

Volcaniclastic sediments (Section 3.11) which include the tuffs, are composed of material of volcanic origin, chiefly lava fragments, volcanic glass and crystals. Volcaniclastics are variable in colour, although many are green through chlorite replacement. They are often badly weathered at surface outcrop.

## 3.2 Sandstones

Sandstones are composed of five principal ingredients: rock fragments (lithic grains), quartz grains, feldspar grains, matrix and cement. The matrix consists of clay minerals and silt-grade quartz, and in most cases this fine-grained material is deposited along with the sand grains. It can form by the diagenetic breakdown of labile (unstable) grains however, and clay minerals can be precipitated in pores during diagenesis. Cement is precipitated around and between grains, also during diagenesis; common cementing agents are quartz and calcite. Diagenetic hematite frequently stains a sandstone red.

The composition of a sandstone is largely a reflection of the geology and climate of the source area. Some grains and minerals are mechanically and chemically more stable than others. Minerals, in decreasing order of stability, are quartz, muscovite, microcline, orthoclase, plagioclase, hornblende, biotite, pyroxene and olivine. A useful concept is that of compositional maturity: immature sandstones contain many unstable grains (rock fragments, feldspars and mafic minerals), mature sandstones consist of quartz, some feldspar and some rock fragments, while supermature sandstones consist almost entirely of quartz. In general, compositionally immature sandstones are deposited close to the source area, while supermature sandstones arise from long distance transport and much reworking. The minerals present in a sandstone thus depend on the geology of the source area, the degree of weathering there and on the length of the transport path.

**Fig. 3.1** Classification of sandstones. Careful use of a handlens in the field should enable recognition of the main sandstone types: quartz arenite, arkose, litharenite and greywacke (after Pettijohn *et al.*, 1973).

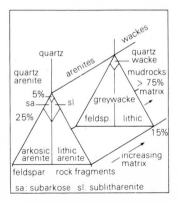

sa: subarkose   sl: sublitharenite

The accepted classification of sandstones is based on the percentages of quartz (+ chert), feldspar, rock fragments and matrix in the rocks (Fig. 3.1). Sandstones containing an additional, non-detrital component are referred to as *hybrid sandstones* and are described in succeeding sections.

The composition of a sandstone is based on a modal analysis, determined from a thin section of the rock using a petrological microscope and a point counter.

In the field, it is often possible to assess the composition and give the sandstone a name, through close scrutiny with a handlens. This can be verified later in the laboratory when a rock slice is available. With a handlens, attempt to estimate the amount of matrix present in a sandstone and thus determine whether it is a wacke or an arenite. The nature of the grains is best determined from their fracture surfaces. The quartz grains will appear milky to clear, glassy, without cleavage surfaces but with fresh conchoidal fractures. Feldspar grains are often slightly (to totally) replaced by clay minerals, so do not have such a fresh glassy appearance as quartz; cleavage surfaces and/or twin planes are usually visible on the fracture surfaces, as they reflect the light. Rock fragments can be recognized by their composite nature and they may show alteration (to chlorite for example). Of the micas, muscovite is recognized by its silvery-grey colour and flaky nature and the less common biotite by its brown-black colour. Some cements in arenites can be identified in the field. Apart from the acid test for calcite, many such cements are large poikilotopic crystals enclosing several sand grains; the cleavage fracture surfaces of such crystals are easily seen with a handlens. Quartz cement usually takes the form of overgrowths on quartz grains. Such overgrowths frequently develop crystal faces and terminations, and these can be seen with a handlens.

## 3.2.1 Quartz arenites

Compositionally supermature, these sandstones are typical of, but not restricted to, high-energy shallow marine environments, and also aeolian (wind-blown) sand-seas in deserts. Sedimentary structures are common, especially cross-stratification, on small, medium and large scales (Chapter 5). Since only quartz is present, the colour of quartz arenites is frequently white or pale grey (especially those of shallow marine environments). Aeolian quartz arenites are frequently red through the presence of finely-disseminated hematite which coats grains. Quartz and calcite cements are common.

Quartz arenites also form through leaching of a sediment whereby all unstable grains are dissolved out. Ganisters, which form in this way, occur beneath coal seams and contain rootlets (black organic streaks).

## 3.2.2 Arkoses

Arkoses can be recognized by the high percentage of feldspar grains although at surface outcrop they may be altered, especially to kaolinite (a white clay mineral). Many arkoses are red or pink, in part due to the presence of pink feldspars but also through hematite pigmentation. Some coarse-grained arkoses look like granites until you see the bedding. In many, grains are subangular to subrounded and sorting is mode-

rate: a considerable amount of matrix may be present between grains. Relatively rapid erosion and deposition under a semi-arid climate produce many arkoses. Non-marine, often fluviatile, environments are typical.

## 3.2.3 Litharenites

Litharenites are very variable in composition and appearance, depending largely on the types of rock fragment present. In phyllarenites, fragments of argillaceous sedimentary rock are dominant and in the calclithites, limestone fragments predominate. Lithic grains of igneous and metamorphic origin are common in certain litharenites. In the field it is usually sufficient to identify a rock as being a litharenite; a more precise classification would have to come from a petrographic study. Many litharenites are deltaic and fluviatile sediments, but they can be deposited in any environment.

## 3.2.4 Greywackes

Greywackes are mostly hard, dark grey rocks with abundant matrix. Feldspar and lithic grains are common and often clearly identifiable with a handlens. Although greywackes are not environmentally restricted, many were deposited by turbidity currents in relatively deep water basins and so show sedimentary structures typical of turbidites (sole structures, graded bedding and internal laminations, Chapter 5).

## 3.2.5 Hybrid sandstones

These contain one or more components which are not detrital, such as the authigenic mineral glauconite or grains of calcite. The *greensands* contain granules of glauconite (a potassium iron aluminosilicate) in addition to a variable quantity of siliciclastic sand grains. Glauconite tends to form in marine shelf environments starved of sediment. *Calcarenaceous sandstones* in particular contain a significant quantity (10—50%) of carbonate grains, skeletal fragments and ooids. With more than 50% carbonate grains, the rock becomes a sandy (or quartzose) limestone. In *calcareous sandstones* the $CaCO_3$ is present as the cement.

For further information on the composition and mineralogy of a sandstone it is necessary to collect samples and study the thin sections made from them. *Petrofacies*, that is sandstones distinct on petrographic grounds, can be of great importance in unravelling the source of the sediment and the palaeogeography at the time.

## 3.3 Conglomerates and breccias

The key features which are important in the description of these sediments are the type of clasts present and the texture of the rock.

On the basis of clast origin, intraformational and extraformational conglomerates and breccias are distinguished. *Intraformational* clasts are derived from within the basin of deposition and many of these are fragments of mudrock or micritic limestone liberated by penecontemporaneous erosion or by desiccation along a shoreline with subsequent re

working (Fig. 3.4). *Extraformational* clasts are derived from outside the basin of deposition and are thus older than the enclosing sediment. The variety of clasts in a conglomerate should be examined: *polymictic conglomerates* are those with many different types of clast, *oligomictic* or *monomictic* conglomerates are those with just one type of clast.

The nature of the extraformational clasts in a conglomerate or breccia is important since it can give useful information on the provenance of the deposit, and on the rocks exposed there at the time. For a meaningful *pebble count*, several hundreds should be identified but you may have to make do with less. If possible undertake pebble counts on conglomerates from different levels in the stratigraphic sequence, and on conglomerates from different parts of the region being studied. These data could show that there were changes in the nature of rocks exposed in the source area during sedimentation, as a result of uplift and erosion, or that several different areas were supplying the material.

For interpretations of the depositional mechanisms of pebbly-bouldery sedimentary rocks, the texture is important: clast-supported conglomerates must be distinguished from matrix-supported conglomerates (Sections 4.4 and 4.6). The shape, size and orientation of the pebbles should be measured, as also the thickness and geometry of the beds and any sedimentary structures.

Conglomerates and breccias are deposited in a range of environments, but particularly glacial, alluvial fan and braided stream. Resedimented conglomerates are those deposited in deep water, usually in a turbidite association.

## 3.4 Mudrocks

Mudrocks are the most abundant of all lithologies but are often difficult to describe in the field because of their fine grain-size. Mudrock is a general term for sediments composed chiefly of silt (4 to 62 μm) and clay (< 4 μm) -sized particles. *Siltstone* and *claystone* are sediments dominated by silt and clay grade material respectively. Claystones can be recognized by their extremely fine grain-size and usually homogeneous appearance; mudrocks containing silt or sand have a 'gritty' feel when crunched between your teeth. *Shales* are characterized by the property of fissility, the ability to split into thin sheets, generally parallel to the bedding; many shales are laminated (Section 5.3.1). *Mudstones* are non-fissile, and many have a blocky or massive texture. *Argillite* refers to a more indurated mudrock while *slate* possesses a cleavage. A *marl* is a calcareous mudrock. Mudrocks grade into sandstones and terms for clay-silt-sand mixtures are given in Fig. 3.2.

Mudrocks are chiefly composed of clay minerals and silt-grade quartz grains: other minerals may be present. Organic matter can be present up to several per cent and higher, and with

**Fig. 3.2** Scheme for describing sedimentary rocks of sand-silt-clay grain sizes (after M. D. Picard).

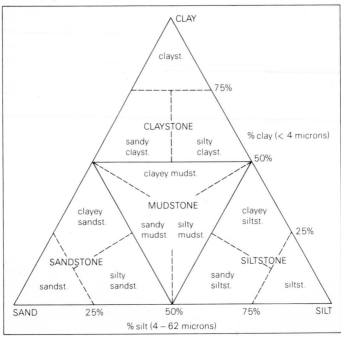

increasing carbon content the mudrock becomes darker and eventually black in colour; a distinctive smell is produced by striking an organic-rich rock with a hammer. Nodules commonly develop in mudrocks, usually of calcite, dolomite, siderite, chert or pyrite (Section 5.5.7). Fossils are present in many mudrocks, including micro-fossils, and need to be extracted in the laboratory.

Mudrocks can be deposited in practically any environment, particularly river floodplain, lake, low energy shoreline, delta, outer marine shelf and deep ocean basin. The sedimentological context of the mudrocks together with the fossil content are important in their environmental interpretation.

In the field, once the type of mudrock present has been ascertained, it can be described by the use of one or two adjectives which relate to a conspicuous or typical feature. Features to note are the colour, degree of fissility, sedimentary structures and mineral, organic or fossil content (Table 3.2).

## 3.5 Limestones

Limestones, like sandstones, can only be described in a limited way in the field; the details are revealed through studies of thin sections and peels. Three components make up the majority of limestones: carbonate grains (allochems); micrite (microcrystalline calcite) and sparite (the cement). The principal allochems are: skeletal grains, ooids, peloids and intraclasts. Many limestones are directly analogous to sandstones, consisting of sand-sized carbonate grains which were moved around on the seafloor, while others can be compared with mudrocks, being fine-grained and composed of lithified carbonate mud (i.e., micrite). Some limestones formed *in situ* by the growth of carbonate skeletons as in reef limestones (Section 3.5.3), or through trapping and binding of sediment by algal mats as in stromatolites (Section 5.4.4).

Modern carbonate sediments are composed of aragonite, high Mg calcite and low Mg calcite. Limestones,

Table 3.2 Features to note and look for when examining mudrocks and examples of terms which can be used in their description.

| Mudrock features | Possibilities and descriptive terms |
|---|---|
| A Note the colour | e.g., grey, red, green, variegated, mottled, etc. |
| B See how the mudrock falls apart | e.g., fissile (shale), non-fissile (mudstone), blocky, earthy, flaggy, papery, cleaved (slate). |
| C Look for sedimentary structures | e.g., bedded or laminated, bioturbated, or massive (apparently structureless). |
| D Check non-clay minerals present | e.g., quartzitic, micaceous, calcareous, gypsiferous, pyritic, sideritic, etc. |
| E Assess the organic content | e.g., organic-rich, bituminous, carbonaceous, organic-free. |
| F Look for fossils | e.g., fossiliferous, graptolitic, ostracod. |

composed of low Mg calcite, form through diagenetic replacement of original aragonite grains and loss of Mg from originally high Mg calcite. Other diagenetic changes important in limestones are dolomitization and silicification.

Although the majority of carbonate sequences are shallow marine in origin (supratidal to shallow subtidal), limestones are also deposited in deeper water as pelagic and turbidite beds, and in lakes. Nodular limestones, called calcretes or caliches, can develop in soils (Section 5.5.7).

### 3.5.1 Limestone components

*Skeletal grains* (bioclasts) are the dominant constituents of many Phanerozoic limestones. The types of skeletal grain present depend on environmental factors during sedimentation (water temperature, depth, salinity, for example) as well as on the state of invertebrate evolution and diversity at the time. The main organism groups contributing skeletal material are the molluscs, the brachiopods, the corals, the echinoderms (especially the crinoids), the bryozoans, the algae and foraminifera. Other groups of lesser or local importance are the sponges, crustaceans (ostracods especially), annelids and cricoconarids.

In the field, one should try and identify the main types of carbonate skeleton in the limestone. If these are present as macro-fossils, it should be possible to identify them to the group level. The carbonate skeletons may be sufficiently well-preserved or abundant to allow useful palaeoecological observations to be made (Chapter 6). One feature to check is whether the skeletal material is in growth posi-

**Fig. 3.3** Ooids; the concentric structure is clearly visible in some cases. Scale in millimetres. Jurassic, N. England.

tion, and if it is, whether this was providing a framework for the limestone during sedimentation, as in reefs (Section 3.5.3).

*Ooids* are spherical to subspherical grains, generally in the range 0.2 mm to 0.5 mm, but reaching several millimetres. Structures larger than 2 mm are referred to as pisoids or pisolites. Ooids consist of concentric coatings around a nucleus, usually a carbonate particle or quartz grain (Fig. 3.3). Most modern ooids are composed of aragonite but all ancient ones are calcitic (unless dolomitized).

*Peloids* are subspherical to elongate grains generally less than 1 mm in

**Fig. 3.4** Intraformational conglomerate consisting of flakes (intraclasts) of micritic limestone. Late Precambrian, Norway.

length. They consist of micrite (see below) and many are faecal in origin.

*Intraclasts* are fragments of re-worked carbonate sediment. Many are flakes up to several centimetres long, derived from desiccation of tidal flat carbonate muds or penecon-temporaneous erosion, and forming intraformational conglomerates (Fig. 3.4). *Aggregates* consist of several carbonate grains cemented together during sedimentation.

*Micrite* is the matrix to many bioclastic limestones and it is the main constituent of fine-grained limestones. It consists of carbonate particles mostly less than 4 μm in diameter. Much modern carbonate mud, the forerunner of micrite, is biogenic in origin, forming through the disintegration of carbonate skeletons such as calcareous algae. The origin of micrite in ancient limestones is obscure and it is often difficult to

**Table 3.3** Schemes for the classification of limestones (*A*) on dominant grain size, (*B*) on dominant constituent; prefixes can be combined if necessary, as in bio-oosparite (after R.L. Folk) and (*C*) on texture (after R.J. Dunham)

| **A** | 2mm | | 62μm |
|---|---|---|---|
| *Calcirudite* | | *Calcarenite* | *Calcilutite* |

| **B** Dominant constituent | *Rock type* | |
|---|---|---|
| | *Sparite cement* | *Micrite matrix* |
| oöids | oosparite | oomicrite |
| peloids | pelsparite | pelmicrite |
| bioclasts | biosparite | biomicrite |
| intraclasts | intrasparite | intramicrite |
| in situ growth : biolithite | | |

| **C** Textural features | | | Rock types |
|---|---|---|---|
| mud absent | grain supported | | grainstone |
| | | | packstone |
| carbonate mud present | mud supported | > 10% grains | wackestone |
| | | < 10% grains | mudstone |
| components organically bound during deposition: | | | boundstone |

eliminate direct or indirect inorganic precipitation.

*Sparite* is a clear equant calcite cement precipitated in the pore space between grains and in larger cavity structures. Fibrous calcite cements also occur, coating grains and lining cavities.

### 3.5.2 Limestone types

Three schemes are currently used for the description of limestones (Table 3.3). Common limestone types using the Folk notation are biosparite, biomicrite, oosparite, pelsparite and pelmicrite. Biolithite refers to a limestone which has formed through *in situ* growth of carbonate organisms (as in a reef). Common limestones, using the R.J. Dunham notation, are grainstones, packstones, wackestones and mudstones; the term boundstone is equivalent to biolithite.

In the field, careful use of a hand-lens can establish the type of limestone present, using either the Folk or Dunham naming system. The main types of grain can be recognized without difficulty, although in finer-grained calcarenites it may be impossible to distinguish between matrix and cement. Next examine the limestone's texture (Chapter 4), bearing in mind that the size, shape, roundness and sorting of skeletal grains in a carbonate sediment is a reflection of the size and shape of the original skeletons as well as the degree of agitation in the environment. Although practically all sedimentary structures of siliciclastics can occur in limestones, there are some which are restricted to carbonate sediments (Section 5.4).

### 3.5.3 Reef limestones

These are *in situ* accumulations or build-ups of carbonate material. Reef limestones have two distinctive features: a massive unbedded appearance (Figs. 3.5 and 3.6), and a dominance of carbonate skeletons, especially of colonial organisms, with many in their growth position. Some skeletons may have provided a framework within which and upon which other organisms existed. Cavity structures infilled with internal sediment and cement are

**Fig. 3.5** A reef complex (Devonian, Canning basin, Australia). Massive, unbedded reef limestones occur in the central part, with well-bedded back reef limestones to the right, and fore-reef limestones to the left, showing an original depositional dip.

**Fig. 3.6** Small patch reef consisting of massive coral colonies and contrasting with bedded bioclastic limestones below. Jurassic, N.E. England.

common. Reef limestones have a variety of geometries but two common forms are: *patch reef*, small and discrete structures (Fig. 3.6), circular in plan, and *barrier reef*, a larger elongate structure with lagoonal limestones behind. *Bioherm* refers to a local carbonate build-up and *biostrome* to a laterally extensive build-up, both with or without a framework. Associated with many reef limestones are beds of reef-derived debris. With barrier reefs especially, reef debris forms a talus apron in front (fore-reef limestones, Fig. 3.5).

One further type of carbonate build-up is the *mud mound* (formerly reef knoll) consisting of massive micrite, usually with scattered skeletal debris, together with cavity structures such as Stromatactis (Section 5.4.1).

With all carbonate build-ups, it is

the massive nature which will be immediately apparent, contrasting with adjacent or overlying well-bedded limestones. Many carbonate build-ups are the result of complex interactions between organisms, so verify which organisms were responsible for the build-up's construction, which were playing a secondary but still important role of encrusting or binding the framework, and which were simply using the reef for shelter or as a source of food.

### 3.5.4 Dolomites

The majority of dolomites, especially those of the Phanerozoic, have formed by replacement of limestones. This dolomitization can take place soon after deposition, i.e., penecontemporaneously, upon high intertidal-supratidal flats of semi-arid regions, or much later during diagenesis. For facies analysis it is important to try and decide which type of dolomite is present. Early dolomite is typically very fine grained and is associated with structures indicative of supratidal conditions: desiccation cracks (Section 5.3.6), evaporites and their pseudomorphs (Section 3.6), cryptalgal laminites (Section 5.4.4) and fenestrae (Section 5.4.1b).

Late diagenetic dolomitization can vary from local replacement of certain grains to wholesale replacement of the rock. There is often an obliteration of the original structure of the limestone through late-stage dolomitization, so that fossils are poorly preserved and sedimentary structures ill defined. On the degree of dolomitization, carbonate rocks can be divided into four categories (Fig. 3.7). Dolomites can possess a higher porosity than the original limestones. Many Precambrian dolomites show no evidence of replacement and may be of primary origin. They show all the features of limestones, with stromatolites (Section 5.4.4) especially common.

### 3.5.5 Dedolomites

Dedolomites are limestones formed by the replacement of dolomite.

**Fig. 3.7** Scheme for dividing limestone-dolomite rocks on increasing dolomite content.

| limestone | 10 | 50 | 90 | dolomite |
|---|---|---|---|---|
| | | dolomitic limestone | calcitic dolomite | |
| 0 | | | | 100% |

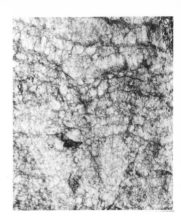

**Fig. 3.8** Nodular (chicken-wire) and bedded gypsum (after anhydrite). Thin clay seams occur between nodules. Tertiary, Iraq. Field of view 2m.

Although relatively rare, they are characterized by unusual growth forms of calcite such as large radiating-fibrous calcite masses ('cannonballs').

## 3.6 Evaporites

Most gypsum occurs as nodular masses within mudrocks or as closely-packed nodules with thin stringers of sediment between (chicken-wire texture) (Fig. 3.8). Layers and beds of gypsum occur, and these may be contorted in the so-called enterolithic texture. Nodular and enterolithic textures are typical of gypsum-anhydrite precipitated in a sabkha (supratidal) environment. Gypsum-anhydrite, interlaminated with organic matter or calcite, is another type, of subaqueous (deeper water) origin. Gypsum-anhydrite can be reworked to display current structures and re-sedimented to form turbidites and slumps. Veins of satin spar are fre-

**Fig. 3.9** Halite pseudomorphs (recognized by the cubic shape and depressed (hopper) crystal faces. Scale in millimetres. Jurassic, S. England.

**Fig. 3.10** Calcite pseudomorphs after gypsum (recognized by the lenticular shape). Scale in millimetres. Jurassic, S. England.

quently associated with gypsum deposits.

Pseudomorphs of evaporites can be recognized in the field but confirmation may require thin sections. Halite pseudomorphs are readily identified by their cubic shape and hopper crystal form (Fig. 3.9). The lozenge, lenticular and swallow-tail shapes of gypsum crystals are distinctive (Fig. 3.10). Nodules of anhydrite

and gypsum can be replaced by a variety of minerals: calcite, quartz and dolomite in particular.

Where evaporites have been dissolved away and not replaced, then collapse of overlying strata has often taken place. Where disrupted and brecciated horizons occur, a careful search may reveal evidence for the former presence of evaporites. Dedolomites are often associated with evaporite solution horizons.

## 3.7 Ironstones

A great variety of sedimentary rocks is included under the ironstone title and there is much variety too in the minerals present. Precambrian banded iron-formations are often thick and laterally-extensive deposits characterized by a fine chert-iron mineral banding. Phanerozoic ironstones are mostly thin sequences of limited areal extent, interdigitating with normal marine sediments. Many such ironstones are oolitic and ooids can be composed of hematite (red), chamosite (green), goethite (brown) and, rarely, magnetite (black). Other common ironstones are hematitic limestones, where hematite has impregnated and replaced carbonate grains, and chamositic, sideritic and pyritic mudrocks. All these types can be identified in the field, but later confirmation is necessary in the laboratory. The iron minerals siderite and pyrite commonly form nodules in mudrocks and other lithologies.

With ironstones, interest has focused on the depositional environment and it is worth examining any fossils contained in the iron-rich beds or in adjacent strata. Check whether the fossils indicate normal marine

31

conditions; they may indicate hypo-saline conditions. Otherwise treat the ironstone like any other lithology and look at its texture and sedimentary structures.

Iron (and other metals) are enriched in sediments associated with pillow lavas. Ferromanganese nodules occur in pelagic limestones and mudrocks but these are rare.

**Fig. 3.11** Bedded chert with shaley partings. Carboniferous, S. France.

## 3.8 Cherts

Two varieties of chert are distinguished: bedded and nodular (Figs. 3.11 and 3.12). Most bedded cherts are found in relatively deep-water sequences and are equivalent to the radiolarian and diatom oozes of the modern ocean floors. With a hand-lens you can see the radiolarians in some bedded cherts, as minute specks (around ¼ mm across); a thin section is required to check their presence. Although many beds of chert appear massive they can possess cross-lamination and graded bedding (Chapter 5). Some bedded cherts are associated with pillow lavas and are part of ophiolite suites, while others occur in sequences with no volcanic associations at all.

Nodular cherts are common in limestones and some other lithologies and form by diagenetic replacement. In some cases there is a nucleus

**Fig. 3.12** Nodular chert (flint), formed within crustacean burrow fills, in chalk. Cretaceous, N.E.England.

around which replacement has proceeded, in others they occur regularly at particular horizons. Flint is a popular name for chert nodules occurring in Cretaceous chalks. Silica for many chert nodules is derived from dissolution of sponge spicules.

## 3.9 Sedimentary phosphate deposits

The phosphate in these relatively rare deposits is mostly collophane, present as vertebrate bone fragments and as nodules. The nodules are often coprolites but in marine phosphorites the nodules may form by replacement of carbonate mud, grains and siliceous microfossils. Reworking is important in the formation of many sedimentary phosphate deposits.

## 3.10 Organic-rich deposits

Peat, brown coal (lignite), hard coal and oil shale are the main organic deposits. Bitumen and other semi-solid/solid hydrocarbons rarely occur within other lithologies. The organic deposits are divided into a humic group, those formed through *in situ* organic growth, chiefly in swamps, and a sapropelic group, where the organic matter has been transported or deposited from suspension. Most coals belong to the humic group whereas oil shales are sapropelic in origin.

The term *rank* refers to the level of organic metamorphism of a coal: a number of properties, such as carbon and volatile content, can be used to measure rank, but these require a laboratory analysis.

In the field, plant material is still recognizable in soft brown coal; with hard brown coal there are few plant fragments visible but the coal is still relatively soft, dull and brown. Brown coals contain much moisture when freshly dug and can be either earthy or compact. Bituminous hard coals are black and hard with bright layers. They break into cuboidal fragments along the cleat (prominent joint surfaces) and dirty the fingers. Anthracite is bright and lustrous with a concoidal fracture.

Cannel and boghead coals are sapropelic deposits which chiefly accumulated in lakes. They are massive, fine-grained, unlaminated sediments, which possess a concoidal fracture.

Oil shales contain more than a third inorganic material, chiefly clay. They are often finely laminated and can be cut with a knife into thin shavings which curl, like wood shavings.

**Table 3.4** Classification of volcaniclastic grains and sediments on grain-size

| Volcaniclastic grains (tephra) | Volcaniclastic sediment terms |
|---|---|
| bombs-ejected fluid / blocks— ejected solid | agglomerate / volcanic breccia |
| 64 mm | |
| lapilli | lapillistone |
| 2 mm | |
| ash | tuff — vitric / lithic \ crystal |
| .06 mm | |
| dust | |

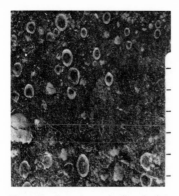

**Fig. 3.13** Accretionary lapilli and ash in lapilli-tuff deposit. Ordovician, N.W. England. Centimetre scale.

## 3.11 Volcaniclastic deposits

Tephra is the term used for material ejected from volcanoes that makes up the volcaniclastic deposits. Tephra include lumps of volcanic glass (pumice or scoria) which fragment to give glass shards; crystals, especially of quartz and feldspar, and lithic fragments of lava from earlier eruptions and of country rock. On the basis of grain-size, the tephra are divided into dust, ash, lapilli blocks and bombs (Table 3.4). Accretionary lapilli consist of an envelope of volcanic dust around a nucleus (Fig. 3.13).

Three types of volcaniclastic deposit are distinguished in terms of origin (Table 3.5). *Pyroclastic-fall deposits* are characterized by a gradual decrease in both bed thickness and grain-size away from the site of eruption. Beds are typically normally graded, can be reworked by currents and waves if deposited in water, or wind if subaerial, and thus may show cross or planar lamination. These deposits can be spread over wide areas and be used for stratigraphic correlation. *Volcaniclastic flow deposits* mostly occur in subaerial situations. Ignimbrites are characterized by an homogeneous appearance with little sorting of the finer ash particles. Coarse lithic clasts may show normal grading and large pumice clasts reverse grading. Flattened and stretched pumice and glass (termed fiamme) demonstrate a flow origin (Fig. 3.14). Many ignimbrites show welding in the central part; here ash particles merge to form a denser, less porous rock compared with the upper and lower parts of the bed. Typical thicknesses of an ignimbrite flow are several metres to ten metres or more. Base-surge deposits are characterized by well-developed internal bedding and cross-stratification (Fig. 3.15), including antidune cross-bedding (Section 5.3.3), since they are deposited by very fast-flowing ash-

**Table 3.5** Main types of volcaniclastic sediment

| |
|---|
| **A**<br>pyroclastic-fall deposits: formed of tephra ejected from vent |
| **B**<br>volcaniclastic flow deposits<br>(and type of flow)<br><br>(i) ignimbrites (fluidized ash-flows)<br>(ii) base surge deposits (ash-laden steam flows)<br>(iii) lahar deposits (mudflows) |
| **C**<br>hyaloclastites: fragmented and granulated basaltic lava through contact with water |

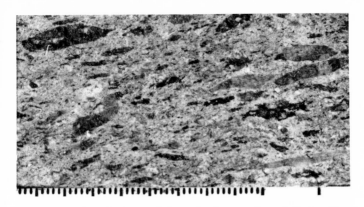

**Fig. 3.14** Ignimbrite showing streaked out glass fragments (fiamme). Tertiary, New Zealand. Millimetre scale.

**Fig. 3.15** Base-surge and pyroclastic fall deposits. The base surge tuff occurs in the upper part and shows well-developed bedding (antidune cross-bedding) with flow to the right. The pyroclastic fall deposits occur in the lower part. Quaternary, Germany. Metre staff.

laden steam flows. Lahar deposits form through volcanic mudflows and are characterized by a matrix-support fabric with 'floating' clasts (Section 4.4). *Hyaloclastites* form where lava is extruded into water and the rapid chilling and quenching causes fragmentation of the lava. These deposits typically consist of lava chips and flakes, a few millimetres to a few centimetres across. They lack any sorting or stratification close to the site of eruption but they can be reworked and resedimented to show sedimentary structures as with any other clastic sediment. Hyaloclastites are typical of submarine basaltic volcanism.

# 4

# *Sedimentary rock texture*

## 4.1 Introduction

Sediment texture is concerned with the grain-size and its distribution, morphology and surface features of grains, and the fabric of the sediment: colour is also considered in this chapter. Texture is useful as a descriptive element of sedimentary rocks and is important in interpreting the depositional mechanism and environment. It is also an important control on the economically significant relationship between porosity and permeability. The texture of many sedimentary rocks can only be studied adequately with rock slices. With sand and silt-sized sediments you cannot do much more in the field than estimate grain-size and comment on the sorting and roundness of grains. With conglom-

erates and breccias grain-size, shape and orientation can be measured accurately in the field; further, surface features of pebbles and the rock's fabric can be examined. A checklist for a sediment's texture is given in Table 4.1.

## 4.2 Sediment grain-size and sorting

The widely-accepted and used grain-size scale is that of Udden-Wentworth (Table 4.2), for detailed work, phi units are used; phi is a logarithmic transformation: $\bar{\phi} = -\log_2 S$, where S is grain-size in millimetres.

For sediments composed of sand-sized particles, use a handlens to determine the dominant grain-size class present; the very coarse, coarse, medium, fine and very fine classes

---

**Table 4.1** Checklist for the field examination of sedimentary rock texture

1 *Grain size and sorting:* estimate in *all* lithologies: Table 4.2 and Fig. 4.1. In conglomerates, measure maximum clast size and bed thickness; check for correlation.

2 *Shape of constituent grains:* Fig. 4.2 (important for clasts in conglomerates); look for facets on pebbles, and striations. Roundness of grains: Fig. 4.3.

3 *Fabric:* (a) look for preferred orientation of elongate clasts in conglomerates and fossils in any lithology; measure orientations and plot rose diagram; (b) look for imbrication of clasts or fossils; and (c) examine matrix–grain relationship, especially in conglomerates, deduce whether matrix-supported or grain-supported sediment.

**Table 4.2** Terms for grain-size classes (after J.A. Udden and C.K. Wentworth) and siliciclastic rock-types. For sand-silt-clay mixtures see Fig. 3.2

|  | boulders | conglomerates (rounded clasts) |
|---|---|---|
| ⊢256mm | cobbles | and |
| ⊢64 | pebbles | breccias |
| ⊢4 | granules | (angular clasts) |
| ⊢2mm | | |
| ⊢1 | v. coarse | |
| ⊢500µm | coarse | |
| SAND | medium | SANDSTONE |
| ⊢250 | fine | |
| ⊢125 | v. fine | |
| ⊢63 microns | | |
| ⊢32 | v. coarse | |
| ⊢16 SILT | coarse | MUDROCKS |
| | SILTSTONE | other types: |
| ⊢8 | medium | mudstone |
| | fine | shale |
| ⊢4 microns | | marl |
| CLAY CLAYSTONE | | slate |

**Table 4.3** Informal terms for describing crystalline rocks

|  |  |
|---|---|
| ⊢2mm | v. coarsely crystalline |
| ⊢1.0 | coarsely crystalline |
| ⊢0.5 . | medium crystalline |
| ⊢0.25 | finely crystalline |
| ⊢0.125 | v. finely crystalline |
| ⊢0.063 | microcrystalline |
| ⊢0.004 | cryptocrystalline |

can be distinguished: silt-grade material feels gritty between the teeth compared with clay-grade material which feels smooth. With chemical rocks such as evaporites, recrystallized limestones and dolomites, it is crystal size that is being estimated, rather than clast size. Terms for crystal size are given in Table 4.3.

For accurate and detailed work, particularly on siliciclastic sediments, various laboratory techniques are available for grain-size analysis, including sieving, point counting on rock slices, and sedimentation methods (see Folk, 1974). In the field only a rough estimate can be made of sorting in a sand grade sediment. Examine the rock with a handlens and compare it with the sketches in Fig. 4.1.

In a broad sense, the grain size of siliciclastic sediments reflects the hydraulic energy of the environment: coarser sediments are transported and deposited by faster-flowing currents than finer sediments; mudrocks tend to accumulate in quieter water. The sorting of a sandstone reflects the depositional process, and this improves with increasing agitation and reworking. In contrast, the grain-size of carbonate sediments generally reflects the size of the organism skeletons and calcified hardparts which make up the sediment; these can be and are affected by currents. Sorting terms can be applied to limestones but bear in mind that some limestone types, oolites and pelleted limestones, for example, are well sorted anyway so that the sorting terms do not necessarily reflect the depositional environment.

For grain-size and sorting of conglomerates and breccias see Section 4.6.

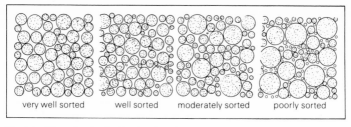

| very well sorted | well sorted | moderately sorted | poorly sorted |

**Fig. 4.1** Charts for visual estimation of sorting.

## 4.3 Grain morphology

The morphology of grains has three aspects: *shape* (or form), determined by various ratios of the long, intermediate and short axes; *sphericity*, a measure of how closely the grain shape approaches that of a sphere; and *roundness*, concerned with the curvature of the corners of the grain. For *shape*, four classes are recognized (Fig. 4.2). These terms are useful for describing clast shape in conglomerates and breccias and can be applied with little difficulty in the field.

Formulae are available for the calculation of sphericity and roundness (see Folk, 1974). Roundness is more significant than sphericity as a descriptive parameter and for most purposes the simple terms of Fig. 4.3 are sufficient. These terms can be applied to grains in sandstones and to pebbles in conglomerates. The terms are less useful for limestones since some grains such as ooids and peloids are well rounded to begin with. Skeletal grains in a limestone should be checked to see if they are broken or their shape has been modified by abrasion.

## 4.4 Sediment fabric

Fabric refers to the mutual arrangements of grains in a sediment. It includes the *orientation* of grains and their *packing*.

In many types of sedimentary rock a *preferred orientation* of elongate particles can be observed. This can be shown by prolate pebbles in a conglomerate or breccia, and fossils in a

**Fig. 4.2** The four classes of grain or clast shape based on the ratios of the long (l), intermediate (i) and short (s) diameters.

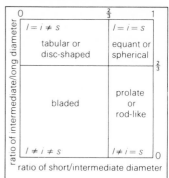

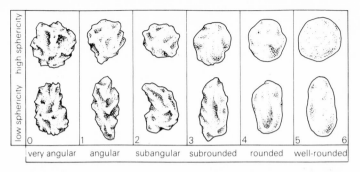

**Fig. 4.3** Categories of roundness for sediment grains. For each category a grain of low and high sphericity is shown.

limestone, mudrock or sandstone; such features are visible in the field. Many sandstones show a preferred orientation of elongate sand grains but microscopic examination is required to demonstrate this.

Preferred orientations of particles arise from interaction with the depositional medium (water, ice, wind), and can be both parallel to (the more common) and normal to flow direction. Measurement of pebble, fossil or grain orientations can thus indicate the palaeocurrent direction (Section 7.3.4). Preferred orientations can also be tectonically induced so if you are working in an area of moderate deformation, also measure fold axes, cleavage and lineations.

Tabular and disc-shaped pebbles or fossils frequently show *imbrication*. In this fabric, they overlap each other (like a pack of cards), dipping in an upstream direction (Fig. 4.4).

The amount of fine-grained matrix and the matrix–grain relationship affect the packing and fabric of a sediment and are important in interpretations of depositional mechanism and environment. Where grains in a sediment are in contact, the sediment is *grain-supported*. Matrix can occur between the grains as can cement. Where the grains are not in contact, the sediment is *matrix-supported* (Figs. 4.4 and 4.5).

With siliciclastic sediments and limestones, grain-support fabric can indicate extensive reworking by currents and/or waves and the removal of mud, or deposition from turbulent flows where suspended sediment (mud) is separated from coarser bed load. Limestones with matrix-support fabric (wackestone and mudstone, Table 3.3) mostly reflect quiet water sedimentation. Fabric of conglomerates and breccias is discussed in Section 4.6.

## 4.5 Textural maturity

The degree of sorting, roundness and matrix content in a sandstone contribute towards the *textural maturity* of the sediment. Texturally immature sandstones are poorly sorted with angular grains and some matrix, while texturally supermature sandstones are well sorted with well-rounded grains and no matrix. Textu-

39

**Fig. 4.4** Conglomerate consisting of rounded to well-rounded pebbles, with a clast-support fabric and well-developed imbrication indicating transport from right to left. Tertiary, Jordan. Field of view 0.75 m.

ral maturity generally increases with the amount of reworking or distance travelled; for example, aeolian and beach sandstones are typically mature to supermature, whereas fluviatile sandstones are less mature. Textural maturity is usually matched by a comparable compositional maturity (Section 3.2). It should be remembered that diagenetic processes can modify depositional texture. An estimate of the textural maturity of a sandstone can be made in the field by close examination with a handlens.

## 4.6 Texture of conglomerates and breccias

There is no problem with measuring the grain-sizes of these coarser sediments in the field; a ruler or tape measure can be used. With conglomerates and breccias, measure the *maximum clast size*, since with many this is a reflection of the competency of the flow. It is also useful to measure the thickness of the conglomerate bed since with some transporting and depositing processes (mudflows and stream floods for example) there is a correlation between maximum particle size and bed thickness. With braided stream conglomerates there is no such relationship. Maximum particle size and bed thickness generally decrease down the transport path. Measurements of maximum particle size and bed thickness from conglomerates over a wide area or from a thick vertical sequence may reveal changes in the supply rates of the sediment and these may reflect either proximity of the source area or more fundamental changes involving climate or tectonics.

For the grain-size distribution in coarse sediments, sorting terms of Fig. 4.1 can be applied but in many cases these terms are inappropriate since the distribution is not uni-modal. Many conglomerates are bi-modal or polymodal in their grain-size distribution if the matrix between pebbles is considered. It is also important to check grain-size variations through a conglomerate bed. Normal size grading of pebbles through a bed is common but reverse grading can also occur, particularly in the basal part (Section 5.3.4).

The shape and roundness of pebbles can be described by reference to Figs. 4.2 and 4.3. With regard to shape, some pebbles of desert and glacial environments possess flat surfaces, *facets* arising either from wind abrasion (such pebbles are known as ventifacts or dreikanters) or glacial abrasion. A characteristic feature of pebbles in a glacial deposit is the presence of striations. Taken over a large area or up a thick sequence,

**Fig. 4.5** Conglomerate with a matrix-support fabric and subangular to subrounded pebbles. Tillite, Late Precambrian, Norway.

there may be significant changes in degree of roundness of pebbles. This can be related to the length of the transport path.

Attention should be given to the fabric of the conglomerate: in particular check for preferred orientations of elongate clasts (if possible measure several tens, or more, of long-axes) and look for imbrication of prolate pebbles (long-axes parallel to current and dipping upstream). If exposures are very good then the dip angle of the long-axis relative to the bedding, can be measured to give the angle of imbrication. In fluvial and other conglomerates a normal-to-current orientation is produced by rolling of pebbles, while the parallel-to-current orientation arises from a sliding of pebbles. In glacial deposits, the orientation of clasts is mostly parallel to the direction of ice movement.

Examine the pebble–matrix relationship (Section 4.4). Pebble-support fabric (Fig. 4.4) is typical of fluvial and beach gravels; matrix-support fabric (Fig. 4.5) is typical of debris flow deposits and glacial tills and tillites.

## 4.7 Colour of sedimentary rocks

Colour can give useful information on lithology, depositional environment and diagenesis. For many purposes a simple estimate of the colour is sufficient although it is amazing how views vary on colour. For detailed work, a colour chart can be used; there are several widely available. It is obviously best to measure the colour of a fresh rock surface but, if different, also note the colour of the weathered surface. The latter can give

an indication of the rock's composition, for example, in terms of iron content.

Two factors determine the colour of many sedimentary rocks: the oxidation state of iron and the content of organic matter. Iron exists in two oxidation states: ferric ($Fe^{3+}$) and ferrous ($Fe^{2+}$). Where ferric iron is present it is frequently as the mineral hematite and even in small concentrations of less than 1% this imparts a red colour to the rock. Where the hydrated forms of ferric oxide, goethite or limonite, are present the sediment has a yellow-brown colour. The formation of hematite requires oxidizing conditions and these are frequently present within sediments of semi-arid and continental environments. Sandstones and mudrocks of these environments (deserts, playa lakes and rivers) are frequently reddened through hematite pigmentation (developed during early diagenesis) and such rocks are referred to as 'red beds'. Many red marine sedimentary rocks are also known. Where reducing conditions prevailed within a sediment the iron is present in a ferrous state and is contained in clay minerals, imparting a green colour to the rock. Green colours can develop through reduction of an originally red sediment, and vice versa. With red and green-coloured deposits see if one colour (usually the green) is restricted to or adjacent to coarser horizons or is concentrated along joint and fault planes, thus indicating later formation.

Organic matter in a sedimentary rock gives rise to grey colours and with increasing organic content a black colour pertains. Organic-rich sediments generally form in anoxic conditions. Finely-disseminated pyrite also gives rise to a dark colour.

Other colours such as olive and yellow can result from a mixing of the colour components. Other minerals with a particular colour in sediments are glauconite and chamosite; they are green.

# 5

# Sedimentary structures and geometry of sedimentary deposits

## 5.1 Introduction

Sedimentary structures are an important attribute of sedimentary rocks. They occur on the upper and lower surfaces of beds as well as within beds. They can be used to deduce the processes and conditions of deposition, the directions of the currents which deposited the sediments (Chapter 7), and in areas of folded rocks, the way-up of the strata. An index of sedimentary structures is given in Table 5.1.

Sedimentary structures are very diverse and many can occur in almost any lithology. Sedimentary structures develop through physical and/or chemical processes before, during and after deposition, and through biogenic processes. It is convenient to recognize four categories of sedimentary structure: erosional, depositional, post-depositional and biogenic.

The geometry of sedimentary deposits is an important feature at all scales and is discussed in Section 5.7.

## 5.2 Erosional structures

The common structures of this group are the flute, groove and tool marks which occur on the undersurfaces (soles) of some sandstone and other beds, scour structures in general and channels.

### 5.2.1 Flute casts

Flute casts are readily identifiable from their shape (Fig. 5.1). In plan, on the bedding undersurface, they are elongate to triangular ('heel-shaped') with either a rounded or pointed upstream end, flaring in a downstream direction. In section they are asymmetric, with the deeper part at the upstream end. Flute marks vary in length from millimetres to tens of centimetres. Flute marks form through local scouring by eddies in currents and are typical of turbidite beds. They also occur on the underside of fluviatile and other sandstones.

Flute marks are reliable indicators of palaeocurrent direction; their orientation should be measured (Chapter 7).

### 5.2.2 Groove casts

Groove casts are elongate ridges on bed undersurfaces, ranging in width from a few millimetres to several tens of centimetres (Fig. 5.2). They may fade out laterally, after several

**Table 5.1** Index of sedimentary structures: main types and location of description and/or figures in this book

| *bedding surface structures* | *bedding undersurface (sole) structures* |
|---|---|
| *ripples:* look at symmetry/asymmetry and crest shape; current, wave or wind ripples? Section 5.3.2, Fig. 5.5, 5.6, 5.7. 5.8 | *flute casts:* triangular, asymmetric structures. Section 5.2.1, Fig. 5.1 |
| *shrinkage cracks:* desiccation or synaeresis cracks: Section 5.3.6, Fig. 5.21 | *groove casts:* continuous/discontinuous ridges. Section 5.2.2, Fig. 5.2 |
| *parting lineation* (primary current lineation) Section 5.3.1, Fig. 5.4 | *tool marks:* various types, Section 5.2.3 |
| *rainspot impressions:* Section 5.3.7, Fig. 5.22 | *load casts:* bulbous structures. Section 5.5.5, Fig. 5.33 |
| *tracks and trails:* crawling, walking, grazing, resting structures. Section 5.6.2, Tables 5.5, 5.6, Figs. 5.39–5.46 | *scours and channels:* small and large scale. Section 5.2.4 and 5.2.5, and Figs. 5.3, 5.18 |

| *internal sedimentary structures* | *structures restricted to or predominant in limestones* |
|---|---|
| *bedding and lamination:* Section 5.3.1, Table 5.2 | *cavity structures:* usually infilled with calcite. Section 5.4.1. Geopetal structures (Fig. 5.23), birdseyes (Fig. 5.24); laminoid fenestrae, usually in cryptalgal limestones; Stromatactis: cavity with flat base and irregular roof (Fig. 5.25) |
| *graded bedding:* normal and reverse. Section 5.3.4, Fig. 5.20 | |
| *cross-stratification:* many types, see Table 5.3 and Section 5.3.3 | *stromatolites:* planar laminated sediments, columns, domes, biostromes, bioherms. Section 5.4.4, Figs. 5.28–5.30 |
| *massive bedding:* Section 5.3.5 | |
| *slumps and slumped bedding:* Section 5.5.1, Fig. 5.32 | |
| *deformed bedding:* various specific types, Section 5.5.2 | *hardgrounds:* recognized by encrusted and bored surfaces. Section 5.4.2, Fig. 5.26 |
| *sandstone dykes:* Section 5.5.3 | |
| *dish structures:* concave-up laminae, pillars between. Section 5.5.4 | *tepees:* pseudoanticlinal structures. Section 5.4.2, Fig. 5.27 |
| *nodules:* Section 5.5.7, Table 5.4 | |
| *stylolites:* sutured planes. Section 5.5.6, Fig. 5.35 | |
| *burrows:* feeding and dwelling biogenic structures. Section 5.6, Figs. 5.39–5.46 | |

metres, or persist across the exposure. Groove casts on a bed undersurface may be all parallel or they may show a variation of trend, up to several tens of degrees or more.

Groove casts form through the infilling of grooves cut chiefly by objects (lumps of mud or wood, etc.) dragged along by a current. Groove casts are common on the undersurfaces of

**Fig. 5.1** Flute marks on undersurface of siliciclastic turbidite. Current flow from left to right. Silurian, N.W. England. Field of view 50 cm.

turbidites and on the soles of other deposits such as fluviatile and storm-deposited sandstones. Groove casts indicate the trend of the current and their orientation should be measured (Chapter 7).

### 5.2.3 Tool Marks

These form when objects being carried by a current come into occasional contact with the sediment surface. The marks are referred to as prod, roll, brush, bounce and skip marks, as appropriate, or simply as tool marks. An impression left by an object may be repeated several times, if it was saltating. Objects making the marks are commonly mud clasts, and fragments of animal and plant debris. As with flutes and grooves, tool marks are seen as casts on the soles of beds, particularly of turbidites.

### 5.2.4 Scour marks and scoured surfaces

These are eroded by currents. Scour marks are small-scale structures, generally less than a metre across, occurring on the base of beds and within them. In plan they are usually elongated in the current direction. With increasing size, scours grade into channels (Section 5.2.5). Typical features of scoured surfaces are the cutting out of underlying sediments, the truncation of underlying laminae and the presence of coarser sediment overlying the scoured surface (Fig. 5.34). The scoured surfaces are often irregular, with some relief, but can be smooth.

Scour marks and scoured surfaces are neither restricted to lithology nor environment but occur wherever currents are sufficiently strong to erode into underlying sediment. They often form during a single erosional event.

### 5.2.5 Channels

Channels are larger-scale structures, metres to kilometres across, that are generally sites of sediment transport over long periods of time. Many channels are concave-up in cross-section (Fig. 5.3) and their fills form elongate

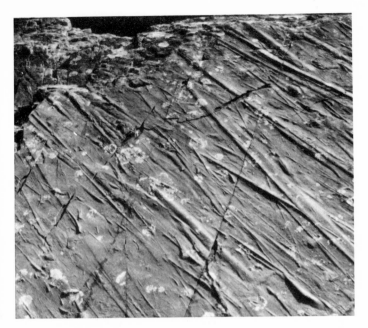

**Fig. 5.2** Groove marks on undersurface of siliciclastic turbidite. Carboniferous, S.W. England. Field of view 2m.

(shoe-string) sediment bodies when mapped in plan. As with scours, channels can be recognized by their cross-cutting relationship to underlying sediments (Figs. 5.3 and 5.18). Channels are frequently infilled with coarser sediment than that below or adjacent, and there is often a basal conglomeratic layer. Cross-bedded sandstones infill many channels.

Some large channels may not be immediately apparent: therefore, view quarry faces and cliffs from a distance and take careful note of the lateral persistence of sedimentary units.

Channels are present in sequences of many different environments, including fluviatile, deltaic, shallow subtidal-intertidal, and submarine fan. With fluviatile and deltaic channels, check for evidence of lateral accretion (Section 5.3.3j) which would indicate lateral migration (meandering) of the channel (Fig. 5.18). Try to measure the orientation of the channel structure; this usually indicates the direction of the palaeo-slope, important in palaeogeographic reconstructions.

## 5.3 Depositional structures

In this group are the familiar structures, bedding, lamination, cross-stratification, ripples and mudcracks. Depositional structures occur on the upper surface of beds and within them. In limestones additional structures are frequently present, including various types of cavity, algal stromatolites, features produced by synsedimentary cementation (hardgrounds and tepees) and subaerial solution (palaeokarstic surfaces).

### 5.3.1 Bedding and lamination

Bedding and lamination define stratification. Bedding is thicker than 1 cm whereas lamination is thinner than 1 cm. Parallel (also called planar or horizontal) lamination is a common internal structure of beds. Descriptive terms for bed and lamination thickness are given in Table 5.2.

#### 5.3.1a Bedding is produced by changes in the pattern of sedimentation; it may be defined by changes in

**Table 5.2** Terminology of bed thickness.

| | |
|---|---|
| —1 metre | very thickly bedded |
| —0.3 m | thickly bedded |
| —0.1 m | medium bedded |
| —0.03 m | thinly bedded |
| —10 mm | very thinly bedded |
| —3 mm | thickly laminated |
| | thinly laminated |

sediment grain-size, colour or mineralogy-composition. Bedding planes can represent long or short breaks in sedimentation; bed junctions can be gradational. Check for evidence of erosion (scour) at bed boundaries, examine bedding planes for such structures as ripples and mudcracks and look at bed undersurfaces for

**Fig. 5.3** Fluvial channel infilled with coarse sand, cutting down into finer sandstones and mudrocks. The cliff is cut perpendicular to the long axis of the channel. Permian, S. Africa.

erosional structures such as casts of flutes, grooves and tool marks. Also examine bed cross-sections for internal sedimentary structures such as cross-stratification and graded bedding. With limestones, bedding planes can be palaeokarstic surfaces, denoting emergence, or hardground surfaces produced through synsedimentary cementation (Sections 5.4.2 and 5.4.3). Be aware, however, that bedding planes, particularly those in limestones and dolomites, can be modified by pressure solution to give stylolites (Section 5.5.6). Bed boundaries can be deformed by compaction and loading; the contacts between sandstones and underlying mudrocks are commonly affected in this way (Section 5.5.5). Tectonic movements, bedding plane slip for example, and the formation of cleavage, can also modify bed junctions. *Bed thickness* is a necessary and useful parameter to measure and in some cases it is not just descriptive. In some conglomerates, for example, there is a relationship between bed thickness and sediment grain-size (Section 4.6). With some current-deposited sediments, turbidites for example, bed thickness decreases in a down-current direction.

*5.3.1b Parallel lamination* (also termed flat bedding) is also defined by grain-size, mineralogical compositional or colour changes. Parallel lamination can be produced in several ways but two basic types are: those formed through deposition from strong currents, referred to as upper plane-bed phase lamination, and those formed through deposition from suspension, low-density turbidity currents or from weak traction currents, the last being referred to as lower plane-bed phase lamination.

*Upper plane-bed phase lamination* predominantly occurs in sandstones and forms through subaqueous deposition at high flow velocities in the upper flow regime (see sedimentology texts). Laminae are several millimetres thick and are made visible by subtle grain-size changes. This type of parallel lamination is characterized by the presence of a *parting lineation*, also called *primary current lineation*, on lamina surfaces (Fig. 5.4). Such surfaces, when seen in the right 'light', have a visible fabric consisting of low ridges only several grain diameters high. This lineation is produced by turbulent eddies close to the sediment surface. Parting lineation is formed parallel to the flow direction, so its orientation will indicate the trend of the palaeocurrent.

*Lower plane-bed phase lamination* lacks parting lineation and occurs in sediments with a grain-size coarser than 0.6 mm. It forms through the movement of sediment as bed load, by traction currents at low flow velocity in the lower flow regime.

**Fig. 5.4** Parting lineation (or primary current lineation), trending from top to bottom of photograph. Devonian, Belgium. Field of view 40cm.

Lamination formed largely by deposition from suspension or low-density turbidity currents occurs in a wide range of fine-grained lithologies, but especially mudrocks and some limestones. Laminae are typically normally graded (Section 5.3.4) if deposited from suspension currents, as is the case with varved sediments of glacial and non-glacial lakes. Laminae can arise from the periodic precipitation of minerals such as calcite, halite or gypsum-anhydrite, and from the blooming of plankton in surface waters with subsequent deposition of organic matter. Many finely-laminated sediments are deposited in protected environments such as lagoons and lakes and in relatively deep-water marine basins below wave-base.

In the field, use a handlens to see the cause of the lamination; is it a fine intercalation of different lithologies or a grain-size change or both? If in a sandstone, split the rock and look for parting lineation on bedding surfaces. If in a limestone, check that the lamination has been formed by the physical movement of grains and that it is not cryptalgal (i.e., stromatolitic) in origin (Section 5.4.4). Measure the average thickness of laminae and the thickness of the parallel-laminated unit.

## 5.3.2 Ripples, dunes and sand waves

These are bedforms developed chiefly in sand-sized sediments. Ripples are common and occur on bedding surfaces, but the larger-scale dunes and sand waves are rarely preserved. The migration of ripples, dunes and sand waves under conditions of net sedimentation gives rise to various types of cross-stratification (Section 5.3.3), which is one of the most common internal depositional sedimentary structures. Both wind and water can move sediment to produce these structures.

*5.3.2a    Wave-formed ripples* are formed by the action of waves on non-cohesive sediment, especially the medium silt to sand grades, and are typically symmetrical in shape; asymmetrical varieties do occur and may be difficult to distinguish from straight-crested current ripples. The crests of wave-formed ripples are generally straight and crest bifurcation is common (Fig. 5.5), sometimes rejoining to enclose small depressions. In profile, the troughs tend to be more rounded than the crests which can be pointed. The ripple index (Fig. 5.7) of wave-formed ripples is generally around 6 or 7. Wave length is affected by sediment grain-size and water depth, larger ripples occurring in coarser sediment and deeper water.

Wave-formed ripples can be affected by changes in water depth to produce *modified ripples*; ripples with flat crests or double crests for example. If there is a change in the direction of water movement over an area of ripples (current or wave-formed), then a secondary set of ripples can develop, producing *interference ripples*. Modified and interference ripples are typical of tidal flat deposits.

*5.3.2b    Current ripples, dunes and sand waves*    Current ripples are produced by unidirectional currents so they are asymmetric with a steep lee side (downstream) and gentle stoss side (Fig. 5.6). On the basis of shape,

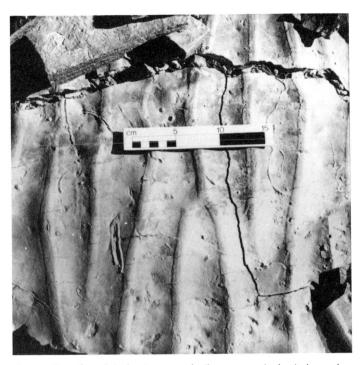

**Fig. 5.5** Wave-formed ripples. Some trace fossils are present in the ripple troughs. Devonian, Libya.

three types of current ripple are common: straight-crested, sinuous or undulatory, and linguoid ripples (Fig. 5.8). Lunate ripples do occur but are rare. With increasing flow velocity of the current, straight-crested ripples pass into linguoid ripples via the transitional sinuous ripples. The ripple index (Fig. 5.7) of current ripples is generally between 8 and 15. Current ripples do not form in sediment coarser than 0.6 mm diameter (coarse sand).

Subaqueous *dunes* (also called megaripples) and *sand waves* (bars) are larger-scale structures of similar

shape to ripples. Although rarely preserved intact the cross-bedding which is produced by their migration is a most common structure (Section 5.3.3b). Dunes are generally a few metres to more than ten metres in length and up to 0.5 m high. Dune shape varies from straight-crested to sinuous to lunate with increasing flow velocity. Ripples frequently occur on the backs and in the troughs of dunes. Sand waves are larger than dunes, being hundreds of metres in length and width, and up to several metres in height. Many are linguoid in shape. Sand waves are present in

**Fig. 5.6** Current ripples. These asymmetric ripples are transitional between straight-crested and linguoid. Current flow from right to left. Triassic, N.W. England.

rivers and similar structures occur on shallow marine shelves. In rivers, sand waves form at lower velocities and at shallower depths than dunes.

*5.3.2c Wind ripples and dunes* are asymmetric like the current ripples. Wind ripples typically have long straight parallel crests with bifurca-tions like wave-formed ripples. The ripple index is high (Fig. 5.7). Wind ripples are rarely preserved. The dunes produced by wind action are also rarely preserved but the cross-stratification produced by their migration is a feature of ancient desert sandstones (Section 5.3.3i). The two common aeolian dune types are

**Fig. 5.7** The ranges of wavelength (L), height (H) and ripple index for wind, wave-formed and current ripples.

| H ←— L —→ | ripple index = L/H |
|---|---|
| wind ripples      L 2.5 – 25 cm   H 0.5 – 1.0 cm | mostly 10 – 70 |
| wave ripples      L 0.9 – 200 cm  H 0.3 – 25 cm | 4 – 13 mostly 6 – 7 |
| current ripples   L < 60 cm   H < 6 cm | > 5  mostly 8 – 15 |

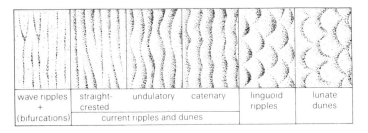

| wave ripples | straight-crested | undulatory | catenary | linguoid ripples | lunate dunes |
|---|---|---|---|---|---|
| (bifurcations) | | current ripples and dunes | | | |

**Fig. 5.8** Crest plans of wave ripples, current ripples and dunes. Linguoid dunes and lunate current ripples are rare. Stoss sides (less steep, upstream facing) are stippled, that is current flowing to right. Dunes are larger scale bed forms than ripples (see text).

barchans (lunate structures) and seifs (elongate sand ridges). The existence of large seif dunes and draas may be revealed by mapping the distribution and thickness of aeolian sandstones over a wide area. Aeolian sandstones have a characteristic large-scale cross-bedding (Section 5.3.3i).

### 5.3.3   Cross-stratification

Cross-stratification is an internal sedimentary structure of many sedimentary rocks and consists of a stratification at an angle to the principal bedding direction. Terms formerly used such as current bedding, false or festoon bedding are best avoided. Much cross-stratification is formed from ripples, dunes and sand waves. However, cross-stratification in sand-grade sediments can also be formed through the infilling of erosional hollows and scours, the growth of small deltas (as into a lake or lagoon), the development of antidunes, the lateral migration of point bars in a channel and deposition on a beach foreshore. Cross-bedding can also form in conglomerates.

Cross-stratification deserves careful observation in the field as it is a most useful structure for sedimentological interpretations, including palaeocurrent analysis (Section 7.3.1). Table 5.3 gives reference to the appropriate section for further information.

*5.3.3a   Cross-lamination and cross-bedding*  Cross-stratification forms either a single *set* or many sets (then termed a *coset*) within one bed (Fig. 5.9). On size alone, the two principal types of cross-stratification are *cross-lamination*, where the set height is less than 6 cm and the thickness of the cross laminae is only a few millimetres, and *cross-bedding* where the set height is generally greater than 6 cm and the individual cross-beds are many millimetres to a centimetre or more in thickness.

*5.3.3b   Shape of cross strata*  Cross-stratification arises from the downstream (or downwind) migration of ripples, dunes and sand waves when sediment is moved up the stoss side and then avalanches down the lee side of the structure. The shape of the cross strata reflects the shape of the lee slope and depends on the charac-

53

**Table 5.3** Cross-stratification in sandstones: how to deal with it

1 Measure (a) set thickness, (b) coset thickness (c) cross bed/cross lamina thickness (d) maximum angle of dip of cross strata (e) direction of dip of cross strata for palaeocurrent analysis (see Chapter 7).

2 Ascertain whether cross lamination (set thickness < 6 cm, cross laminae < a few mm thick) or cross bedding (set thickness usually > 6 cm, cross beds more than few mm thick) (Section 5.5.3a).

3 If cross lamination:–
    (a) examine shape of foresets: tabular or trough (Section 5.3.3b, Fig. 5.12).
    (b) is it climbing-ripple cross lamination (Section 5.3.3d), are stoss sides erosional surfaces or are stoss-side laminae preserved (Fig. 5.13)?
    (c) is it current-ripple or wave-ripple cross lamination? Look for form-discordant laminae, draping foreset laminae, undulating or chevron lamination all of which characterize wave-formed cross lamination (Section 5.3.3e, Fig. 5.14)
    (d) are there mud drapes giving flaser bedding or interbedded mud horizons giving wavy bedding or are there cross-laminated lenses in mudrock giving lenticular bedding (Section 5.3.3f, Fig. 5.15)?

4 If cross bedding:–
    (a) examine shape of cross-bed sets: trough, tabular or wedge-shaped (Section 5.3.3b, Fig. 5.12).
    (b) examine foresets: planar or trough beds, angular or tangential basal contact.
    (c) check bottom sets for cross laminae: co-flow or back-flow (Section 5.3.3b, Fig. 5.10).
    (d) examine texture of sediment: note distribution of grain-size, look for sorting and grading of sediment in cross beds and alternations of coarse and fine beds (Section 5.3.3c).
    (e) look for evidence of current reversals in herring-bone cross-bedding (Section 5.3.3g).
    (f) look for internal erosion surfaces within cross-bed sets; are they reactivation surfaces (Section 5.3.3h)?
    (g) look for low-angle surfaces within the cross-bedded unit; are they lateral accretion surfaces (Section 5.3.3i)?
    (h) consider an aeolian origin (features are thick cross-bed sets and high dip angles, Sect. 5.3.3j), a beach-foreshore origin (features are low-angle cross strata in truncated sets, Section 5.3.3k) or formation through progradation of small deltas (Section 5.3.3l).

teristics of the flow, water depth and sediment grain-size. The steeply-dipping parts of the cross strata are referred to as *foresets* and can have either *angular* or *tangential* contacts with the horizontal: in the latter, the lower less steeply dipping parts are called *bottom sets*. Cross-lamination, upstream-directed (back-flow) or downstream-directed (co-flow), can be developed within the bottom sets of large-scale cross-beds as a result of ripple formation in the dune trough (Fig. 5.10). The upper bounding surface of a cross-bed set is invariably an erosion surface; this is the case with most sets of cross-lamination too, but occasionally stoss-side laminae are preserved (Section 5.3.3d).

Where the original bedform pro-

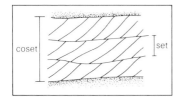

**Fig. 5.9** A coset with three sets of cross-stratification.

ducing the cross-stratification is preserved (usually a ripple) and the cross-stratification is concordant

with this shape, the cross-stratified unit is referred to as a *form set*.

The three-dimensional shape of cross-stratified units defines two common types: *tabular cross strata*, where inter-set boundaries are generally planar, and *trough cross strata*, where the inter-set boundaries are festoon (trough) shaped (Fig. 5.11). Wedge-shaped cross-stratified units also occur (Fig. 5.12). Tabular and wedge cross-bedding mostly consists of planar beds which have an angular contact with the basal surface

**Fig. 5.10** Large-scale cross bedding (set height 60cm) with tangential, concave-up cross-beds, and well-developed bottom sets with cross-lamination. Also note the increase in fine material in the lower part of the foresets as a result of deposition from suspension in the trough of the bedform. Current flow from left to right. Deformed bedding in the upper part is a result of sediment dewatering. Carboniferous, N.E. England.

**Fig. 5.11** Planar cross-beds in wedge-shaped sets, together with pronounced overturning of foresets. Current flow from left to right. Carboniferous, N.E. England.

of the set. Trough cross-bedding consists of scoop-shaped beds, with tangential bases.

Tabular cross-stratification is produced by straight-crested, i.e. two-dimensional, bedforms whereas trough cross-stratification results from curved-crested, i.e. three-dimensional, bedforms. Tabular cross-lamination is formed by straight-crested ripples; tabular cross-bedding is mainly produced by sand waves, and also by straight-crested dunes. Trough cross-lamination is chiefly produced by linguoid ripples and trough cross-bedding is mainly formed by lunate and sinuous dunes.

*5.3.3c Sorting in cross-beds* Close observation of the individual cross-

beds will show variations in the grain-size distribution. Where avalanching of sediment down the lee slope was intermittent, cross-beds show good sorting with coarser particles concentrated towards the outer (down-stream) part and towards the base. Coarser and thicker beds formed by avalanching can alternate with thin beds of fine material deposited from suspension between avalanche events. Where avalanching was continuous then sorting is less well defined and fine beds are absent. Finer sediment and plant debris are often concentrated in bottom sets of cross-beds and cross-laminae, since these are carried over the ripple or dune crest and deposited in the trough.

Sections 5.3.3*d* to 5.3.3*n* give

examples of some particular types of cross-stratification

### 5.3.3d Climbing-ripple cross-lamination

When ripples are migrating and much sediment is being deposited, ripples will climb up the backs of those downstream to form climbing-ripple cross-lamination, also called ripple drift. With rapid sedimentation, stoss-side laminae can be preserved so that laminations are continuous (Fig. 5.13).

### 5.3.3e Wave-generated cross-lamination

The internal structure of wave-formed ripples is variable; often the laminae are not concordant with the ripple profile (i.e. the laminae are form-discordant). Two other features which distinguish wave-formed from current ripple cross-lamination are irregular and undulating lower set boundaries and draping foreset laminae (Fig. 5.14).

### 5.3.3f Flaser, lenticular and wavy bedding

In some areas of ripple formation, the ripples of silt and sand move periodically and mud is deposited at other times. *Flaser bedding* is where cross-lamination contains mud streaks, usually in ripple troughs. *Lenticular bedding* is where mud dominates and the cross-lamination occurs in sand lenses. *Wavy bedding* is where thin ripple cross-laminated beds alternate with mudrock, see Fig. 5.15. These bedding types are common in tidal flat and delta front sediments, wherever there are fluctuations in sediment supply or level of current (or wave) activity.

### 5.3.3g Herring-bone cross-bedding

applies to bipolar cross-bedding, where cross-bed dips of adjacent sets are oriented in opposite directions (Fig. 5.16). Herring-bone cross-bedding is produced where there are reversals of the currents, causing dunes and sand waves to change their direction of migration. It is a characteristic but not ubiquitous feature of tidal sand deposits.

### 5.3.3h Reactivation surfaces

With some cross-bed sets careful observation will show that there are erosion sufaces within them, cutting across the cross-strata (Fig. 5.16). These reactivation sufaces represent short-term changes in the flow conditions which caused modification to the

**Fig. 5.12** Tabular (a) and trough (b) cross-stratification. In (a), cross-beds are generally planar with angular basal contacts, whereas in (b) cross-beds are scoop-shaped with tangential bases.

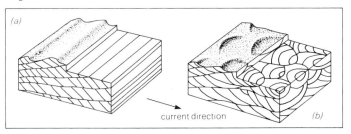

*(a)*

current direction

*(b)*

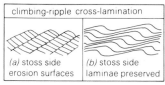

| climbing-ripple cross-lamination |
| --- |

| (a) stoss side erosion surfaces | (b) stoss side laminae preserved |

**Fig. 5.13** Two types of climbing-ripple cross-lamination (ripple drift). In (a), sets of cross-laminae are bounded by erosion surfaces; in (b) stoss side laminae are preserved so that cross-laminae are continuous.

shape of the bedform. They can occur in tidal sand deposits through tidal current reversals (where the current of one direction is of insufficient strength to cause much sediment movement and thus cross-bedding in that direction), in fluviatile sediments through changes in river stage, and in aeolian sands through changes in wind strength.

### 5.3.3i   Aeolian cross-bedding

Compared with cross-bedding of subaqueous origin, cross-bedding produced by wind action generally forms sets which are much thicker; the cross-beds themselves dip at higher angles (Fig. 5.17). Sets of aeolian cross-beds are typically several metres (up to 30 m) in height. Cross-beds can be trough or planar in shape and they most commonly have tangential bases. Foresets often dip at angles in excess of 30°. Aeolian sandstone sequences typically consist solely of large-scale cross-bedded sets. If an aeolian origin is suspected, then also look at the composition and texture of the sediment (Sections 3.2.1 and 4.2). Cross-bedding formed subaqueously is generally less than 2 m in thickness and cross-bed angles of dip are generally less than 25°.

### 5.3.3j   Lateral accretion surfaces/ epsilon cross bedding

Within cross-bedded channel sandstones there can sometimes be discerned a larger-scale cross-bedding, oriented normal to smaller-scale cross-stratification (Fig. 5.18). This *epsilon cross-bedding* forms through lateral migration of the channel and represents the successive growth of point bars. Usually only a few of these lateral accretion surfaces will be present in a field section, spaced at distances of several metres. They are generally a metre or more in height and continue laterally for several metres to more than ten metres. Lateral accretion surfaces are typical of meandering river channel sandstones but they can also occur in delta distributary and tidal channel deposits.

### 5.3.3k   Small-delta cross-bedding

Where small deltas build into lakes

**Fig. 5.14**   Three types of internal structure of wave-formed ripples.

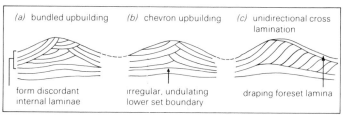

| (a) bundled upbuilding | (b) chevron upbuilding | (c) unidirectional cross lamination |
| --- | --- | --- |
| form discordant internal laminae | irregular, undulating lower set boundary | draping foreset lamina |

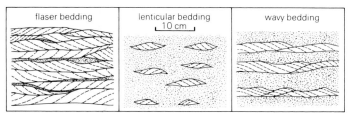

**Fig. 5.15** Sketches of flaser, lenticular and wavy bedding.

and lagoons, large-scale cross-bedding can develop which represents the prograding front of the delta (delta slope). The thickness of the cross-bedded unit (generally a few metres) reflects the depth of water into which the delta was building. Foresets dip at angles of 10° to 25° and consist of sand, passing into much finer-grained and well-developed bottom sets of silt and clay, deposited largely from suspension in front of the delta. Top sets are also well developed and can consist of lenticular gravels, sands and finer sediments, deposited by streams on the delta top. Small-delta cross-bedding is identified by the presence of well-developed top sets and bottom sets (the former contrasting with cross-bedding formed from dunes and sandwaves) and the presence of one thick set; it

occurs as a wedge or fan in marginal lacustrine or lagoonal settings.

*5.3.31 Antidune cross-bedding* is rare but important since it indicates high-flow velocities in the upper flow regime. Antidunes are bedforms in sand-grade sediment which migrate upstream through deposition of sediment on the upstream-facing slope of the bedform. The cross-bedding formed by the migration of antidunes is directed upcurrent, so other evidence of flow direction is required to be certain that the supposed antidune cross-bedding is oriented against the flow. This type of cross-bedding generally has a lower angle of dip and is less well defined than that formed by lower flow velocity subaqueous dunes. Antidune cross-bedding is known from turbidite and fluvial

**Fig. 5.16** Herring-bone cross-bedding and a reactivation surface in cross-bedding.

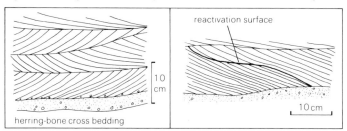

sandstones, and volcaniclastic base surge deposits (Fig. 3.15).

### 5.3.3m Beach cross-stratification

Siliciclastic and carbonate sands deposited on a beach of moderate to high wave activity, are characterized by a low-angle planar cross-stratification arranged in truncated sets (Fig. 5.19). The low angle flat bedding is typically directed offshore, but shoreward-directed bedding also develops through sand deposition on the landward side of a beach berm. Boundaries between sets represent seasonal changes in the beach profile. The lamination is formed by wave swash-backwash and commonly possesses primary current lineation (Section 5.3.1b). The texture and composition of the sediment may help to confirm a beach origin (Sections 4.2 and 3.2.1).

### 5.3.3n Cross-bedding in conglomerates

is not uncommon and is usually in single sets, with set heights in the range of 0.5 m to 2 m. Cross-beds, usually planar, can occur in tabular, wedge and lenticular units. Cross-bedding is common in conglomerates deposited in fluviatile environments (braided streams and stream floods) where it forms from the downstream migration of bars at high stage.

### 5.3.4 Graded beds

These show grain-size changes from bottom to top. The most common is *normal graded bedding* where the coarsest particles at the bottom give way to finer particles higher up (Fig. 5.20). The decrease in grain-size upwards can be shown by all particles in the bed or by the coarsest particles only, with little change in the grain-size of the matrix. *Composite* or multiple-graded bedding is where there are several graded units within one bed.

Less commonly *reverse* (or inverse) grading is developed, where the grain-size increases upwards. This can occur throughout a bed, or more commonly it occurs in the bottom few centimetres of the bed, with normal graded bedding following. Reverse grading may only affect the coarse particles. Graded bedding can be observed (and measured) in conglomerates with no difficulty and in sandstones with the aid of a handlens.

Normal graded bedding often results through deposition from waning flows; as a flow decelerates so the coarsest (heaviest) particles are deposited first and then the finer

**Fig. 5.17** Aeolian cross-bedding characterized by set heights of several to many metres, and high angles of dip (often more than 30°). Height of exposure 15m. Permian, N.E. England.

**Fig. 5.18** Lateral accretion surfaces (arrowed) giving a large scale (epsilon) cross-stratification, seen in a cliff section across small channel, which in fact is cut by a larger channel. Palaeocurrent direction towards observer, as deduced from smaller-scale cross-bedding within the sandstone units and the orientation of the channel-fill itself. The channel-fills are overlain by a coarsening-upward sandstone capped by a thin coal seam. Carboniferous, N.E. England. Height of cliff 7m.

particles. Such graded bedding is typical of turbidity current and storm current deposits. Composite graded bedding can reflect pulses in the current.

Reverse grading can arise from an increasing strength of flow during sedimentation and other effects. Laminae deposited on beaches by backwash are frequently reversely graded and reverse grading can occur in the lowest parts of mass sediment flow deposits such as grain flow and debris flow deposits.

**Fig. 5.19** Beach-foreshore bedding consisting of truncated sets of low-angle parallel-laminated sand (with parting lineation). Units of cross-lamination may occur, together with some burrows and heavy mineral concentrations.

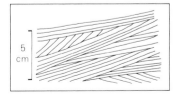

### 5.3.5  Massive beds

Massive beds have no apparent internal structure. It is first necessary to ascertain that this really is the case and that it is not simply due to surface weathering. Blocks collected in the field and cut and polished in the laboratory may show that structures are indeed present. Staining the surface or using X-radiography at the local hospital may bring to light structures.

If the bed really is structureless then this is of interest and attempts should be made to deduce why. Two alternatives are that it was deposited without any structure or that the depositional structure has since been destroyed by such processes as bioturbation (Section 5.6.1), recrystallization, and dewatering (Section 5.5). Where an original structure has been destroyed, careful examination of the bed, or again of cut and stained blocks, may reveal evidence for this; wisps of lamination may remain, the sediment may appear churned and homogenized. Truly massive beds

61

deposited in this condition mostly arise through rapid sedimentation ('dumping'), where there was insufficient time for bedforms to develop. Massive bedding is a feature of some turbidity current and grain flow sandstones, and debris flow deposits.

### 5.3.6 Shrinkage cracks (mud-cracks)

These are present in many fine-grained sediments and most form through desiccation on emergence, causing a shrinkage of the bed or lamina and thus cracking. Many desiccation cracks define a polygonal pattern on the bedding surface (Fig. 5.21). Polygons vary enormously in size, from millimetres to metres across. Several orders of crack pattern may be present. Sediment clasts can be liberated by desiccation.

**Fig. 5.20** Graded bedding in sandstone, from very coarse sand grade and granules at the base to medium sand at the top. Precambrian, Scotland.

Sediments also crack subaqueously: *synaeresis cracks* form through sediment dewatering, often resulting from salinity changes. Synaeresis cracks are characterized by an incomplete polygonal pattern. The cracks are often trilete or spindle-shaped (Fig. 5.21) and can be mistaken for trace fossils.

Desiccation and synaeresis cracks are often infilled with coarser sediment, seen in vertical section as wedges, although these can be deformed and folded through compaction. Desiccation cracks indicate subaerial exposure and so are common in sediments of marine and lacustrine shorelines. Synaeresis cracks are not uncommon in shallow sublittoral lacustrine deposits.

### 5.3.7 Rainspots

Rainspots are small depressions with rims, formed through the impact of rain on the soft exposed surface of fine-grained sediments (Fig. 5.22).

## 5.4 Depositional structures of limestones (including dolomites)

### 5.4.1 Cavity structures

Many limestones contain structures which were originally cavities but were infilled with sediment and/or carbonate cement soon after deposition. These include geopetal structures, fenestrae including birdseyes, stromatactis, sheet cracks and neptunian dykes.

*5.4.1a Geopetal structures* This term can be applied to any cavity (not only in a limestone) infilled with internal sediment and cement (normally

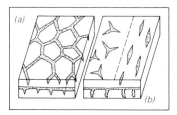

**Fig. 5.21** Shrinkage cracks. (a) formed by desiccation, typically complete polygons, can be straight-sided as shown, or less regular. (b) formed through synaeresis, typically incomplete with either bird's foot or spindle shape. In (a), the cracks are depicted as having suffered little subsequent compaction and so appear V-shaped in section; in (b) the crack infills are ptygmatically folded through compaction.

sparry calcite). The whole cavity infilling is a most useful way-up indicator (the white sparite at the top) and the surface of the internal sediment provides a 'spirit-level', showing the position of the horizontal at or just after deposition. Geopetal structures commonly form beneath and within skeletal grains (Fig. 5.23).

Careful measurement of geopetal structures can show that a series of limestones had an original depositional dip. This is often the case with fore-reef limestones.

**Fig. 5.22** Rainspots on surface of mudstone. Permian, S. Africa.

### 5.4.1b Fenestrae (including birdseyes)
are cavity structures, usually infilled with sparite, occurring in micritic, often pelleted, limestones or dolomites. Two common types of fenestrae are: equant to irregular fenestrae (birdseyes) and laminoid fenestrae. Birdseyes are typically a few millimetres across and are formed through gas entrapment and desiccation in tidal flat carbonate sediments (Fig. 5.24). Laminoid fenestrae are elongate cavity structures, parallel to the stratification, often occurring between cryptalgal laminae.

### 5.4.1c Stromatactis
is another specific type of cavity which is characterized by a smooth floor of internal sediment, an irregular roof and a cement infilling, usually fibrous calcite followed by drusy sparry calcite (Fig. 5.25). Stromatactis is common in mud-mound limestones (massive biomicrites), but its origin is unclear. Sediment dewatering, local seafloor cementation and sediment scouring are likely explanations.

### 5.4.1d Sheet cracks and neptunian dykes
are continuous cavities either parallel to or cutting the bedding. Both can vary considerably in size, particularly the neptunian dykes which can penetrate down many metres. They can be infilled with sediment of either similar age and lithology to the host sediment, or with quite different and much later material. Sheet cracks and neptunian dykes form through penecontemporaneous tectonic movements causing cracking and fissuring of a limestone mass.

Somewhat similar structures, but often on a larger scale, can develop within a limestone mass when it is uplifted and brought into contact with meteoric waters. Solution (karstification) of the limestone can result in cave systems being formed which are subsequently infilled with sediment, usually red and green nonmarine marls and coarser sediments.

## 5.4.2 Hardgrounds and tepee structures

These are limestones showing evidence of synsedimentary subtidal cementation so that the sediment was partly or wholly lithified (hard) on the seafloor. The top surface of the hardground usually provides the best evidence for a cemented seafloor. A hardground surface is usually encrusted by sessile organisms such as oysters, serpulid worms and crinoids, and penetrated by the borings of such organisms as annelids, lithophagid bivalves and sponges (Fig. 5.26). Many hardground surfaces are planar, having formed by corrasion, and can often be traced over considerable areas. Other hardgrounds have a more irregular surface with some relief where a degree of seafloor solution has taken place.

As a result of the synsedimentary cementation of carbonate sediments the cemented surface layer can expand and crack into a polygonal pattern. The cemented crust can be pushed up to form *tepees* (pseudo-anticlines,

**Fig. 5.24** Birdseyes (fenestrae) infilled with sparite within pelleted limestone. Carboniferous, Wales. Millimetre scale.

Fig. 5.27), and where cracked the crust can be thrust over itself. Cavities can develop beneath the cemented surface layer and be infilled with sediment and further cement. Tepee structures can develop in shallow subtidal sediments, in conjunction with hardgrounds, and in tidal-flat carbonates. Hardgrounds and tepee structures are not common but they do give important environmental and diagenetic information.

## 5.4.3 Palaeokarstic surfaces

Palaeokarstic surfaces showing the effects of subaerial solution will have an irregular topography, possibly with potholes, and be overlain by a thin clay bed which may represent a soil; the mineralogy of this clay may be distinctive.

Also associated with palaeokarstic surfaces are *laminated crusts*. These occur on the top of the limestone, although they may be cut by the solution surface. They typically consist of pale brown micritic carbonate with a poorly-defined lamination; small tubes may be present which were the

**Fig. 5.23** Geopetal structure beneath a shell.

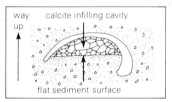

way up

calcite infilling cavity

flat sediment surface

64

**Fig. 5.25** Stromatactis cavities (irregular roof, fibrous and drusy calcite infill, and smooth floor of internal sediment) within biomicrite. Carboniferous, N.W. England.

sites of rootlets. Palaeokarstic surfaces are important since they indicate prolonged periods of subaerial exposure.

### 5.4.4 Stromatolites

Stromatolites are organo-sedimentary structures consisting of laminations which define a variety of growth forms. They develop through the trapping and binding of carbonate particles by a surficial mat mainly composed of blue-green algae. Stromatolites are very common in Precambrian carbonate sequences but also occur in many of Phanerozoic age.

Stromatolites vary from planar laminations, called cryptalgal laminites (Fig. 5.28), often showing the effects of desiccation, to domes (like cabbages) and columns (cigar-shaped). Laminations are less than several millimetres in thickness and consist of micrite, peloids and fine skeletal debris. With planar forms the cryptalgal origin of the laminae is shown by small corrugations and undulations and preferential thickening over small surface irregularities. Elongate cavities (laminoid fenestrae, Section 5.4.1b) are common in cryptalgal laminites.

A useful notation for describing stromatolites in the field is given in Fig. 5.29. The morphology of stromatolites frequently changes upwards, as a response to environmental changes, and large stromatolite structures may consist of lower orders of domes and columns. The notation of

**Fig. 5.26** View of hardground surface showing encrusting oysters and borings (circular holes). Jurassic, W.England.

Fig. 5.29 provides a convenient short-hand for describing such changes and variations.

Oncolites are spherical to subspherical, unattached cryptalgal structures (algal balls), often with concentric laminations (Fig. 5.30).

## 5.5 Post-depositional sedimentary structures

A variety of structures are formed after deposition, some through mass movement of sediment (slumping) and others through internal reorganization by dewatering and loading. Post-depositional physico-chemical and chemical processes produce stylolites, solution seams and nodules.

### 5.5.1 Slumps and slides

Once deposited, either upon or close to a slope, a mass of sediment can be transported downslope. Where there is little internal deformation of the sediment mass, more often the case with limetones, then the transported mass is referred to as a slide. Brecciation of the sediment mass can take place to produce large and small blocks. Where a sediment mass is internally deformed during down-slope movement, then the term slump is more appropriate. A slumped mass typically shows folding; recumbent folds, asymmetric anticlines and synclines and thrust folds are common, on all scales (Fig. 5.31). Fold axes are oriented parallel to the strike of the

**Fig. 5.27** Tepee structure in laminated limestone. Triassic, Wales. Field of view 1 m.

slope and the direction of overturning of folds is downslope. It is thus worth while measuring the orientation of fold axes and axial planes of slump folds to ascertain the direction of slumping and palaeoslope. Slumps and slides can be on the scale of

metres or kilometres. Many are triggered by earthquake shocks.

The presence of a slump or slide in a sequence can be deduced from the occurrence of undisturbed beds above and below, and a lower contact (the surface upon which the slump or slide took place), which cuts across the bedding (Fig. 5.32). It is necessary to convince yourself that lateral mass movement of sediment has taken place; somewhat similar convolutions and brecciations of strata can be produced by dewatering (see below) and other processes.

### 5.5.2 Deformed bedding

Deformed bedding and terms such as disrupted, convolute and contorted bedding can be applied where the bed-

**Fig. 5.28** Cryptalgal laminites. Precambrian, Norway.

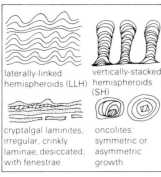

| laterally-linked hemispheroids (LLH) | vertically-stacked hemispheroids (SH) |
| cryptalgal laminites, irregular, crinkly laminae, desiccated, with fenestrae | oncolites: symmetric or asymmetric growth |

**Fig. 5.29** Four common cryptalgal structures, domal, columnar and planar stromatolites, and oncolites.

ding, cross-bedding and cross-lamination produced during sedimentation have been subsequently deformed, but where there has been no large-scale lateral movement of sediment.

*Convolute bedding* typically occurs in cross-laminated sediments, with the lamination deformed into rolls, small anticlines and sharp synclines. Such convolutions are frequently asymmetric or overturned in the palaeocurrent direction. *Contorted* and *disrupted bedding* applies to less regular deformations within a bed, involving irregular folding, contortions and disruptions without any preferred orientation or arrangement. Wholesale or local brecciation of some beds can occur. *Overturned cross-bedding* affects the uppermost part of cross-beds and the overturning is invariably in the downcurrent direction.

Deformed bedding can arise from a number of processes. Shearing by currents on a sediment surface and frictional drag exerted by moving sand are thought to cause some convolute bedding and overturned cross-bedding. Dewatering processes such as fluidization and liquefaction (often

**Fig. 5.30** Oncolites. Carboniferous, N.E. England.

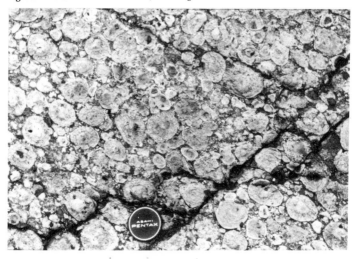

**Fig. 5.31** Slump folded, thrust and brecciated nodular limestone beds. Devonian, Germany. Field of view 1.5m.

induced by earthquake shocks) give rise to convolutions, contortions and disruptions.

### 5.5.3 Sandstone dykes and sand volcanoes

These are relatively rare structures and readily identifiable, the dykes from their cross-cutting relationship with bedding and infilling of sand, and the sand volcanoes (formed where sand moving up a dyke reached the sediment surface) from their conical shape with a central depression, occurring on a bedding plane.

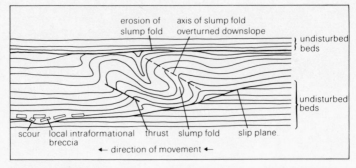

**Fig. 5.32** Principal features of a slumped bed. Slumps can occur on a scale of centimeters to kilometres.

Dewatering, often earthquake-shock initiated, is the cause of these structures.

### 5.5.4 Dish and pillar structures

These consist of concave-up laminae (the dishes), generally a few centimetres across, separated by structureless zones (the pillars). Dish and pillar structures are formed by the lateral and upward passage of water through a sediment. Although not restricted to sandstones of a particular environment or depositional mechanism, they are common structures of sediment flow deposits which occur in submarine fan sequences.

### 5.5.5 Load structures

Load structures are formed through differential sinking of one bed into another. *Load casts* are common on the soles of sandstone overlying mudrock, occurring as bulbous, rounded structures, generally without any preferred elongation or orientation (Fig. 5.33). Mud can be injected up into the sand to form flame structures (Fig. 5.34). Also as a result of

loading, a bed, usually of sand, can sink into an underlying mud and break up into discrete masses, forming the so-called ball and pillow structure. On a smaller scale, individual ripples can sink into underlying mud, producing sunken ripples or sandstone balls.

Close attention should be paid to sandstone–mudrock junctions since it will often be found that gravity-loading has taken place (Fig. 5.34).

### 5.5.6 Pressure solution and compaction

A result of overburden and tectonic pressure is that solution takes place within sedimentary rock masses along certain planes. Pressure solution effects are commonly seen at the junctions of limestone beds and within limestones in the form of *stylolites*, sutured to irregular solution seams with insoluble material (chiefly clay) concentrated along them (Fig. 5.35). Solution seams with less relief occur in more argillaceous limestones and they can pass into cleavage planes in adjacent mudrocks. The occurrence of pressure solution can be demon-

**Fig. 5.33** Load structures on underside of sandstone bed. Late Precambrian, Norway. Field of view 1m.

strated by the partial loss of fossils across stylolites. Extreme pressure solution can give a limestone a brecciated appearance. Stylolites also occur in sandstones and between pebbles in conglomerates.

Compaction of muddy sediments begins soon after deposition and the main effects are the crushing of fossils and reduction of deposited sediment thickness by a factor of up to 10 (best seen where there are early diagenetic nodules in the mudrock: Section 5.5.7). Compaction coupled with pressure solution also enhances limestone–mudrock contacts.

## 5.5.7 Nodules

Nodules, (also called concretions) commonly form in sediments after deposition (Fig. 5.36). Minerals often comprising nodules are fine-grained

varieties of calcite, dolomite, siderite, pyrite, collophane and quartz (chert). Calcite, pyrite and siderite nodules of a few millimetres to a few tens of centimetres in diameter are common in mudrocks; chert nodules are com-

**Fig. 5.34** (a) scoured surface at base of sand bed; truncation of mud laminae at contact. (b) loaded surface, through sinking of sand into mud. Upward injection of mud (flame structure) and depression and contortion of mud laminae. A scoured surface can be deformed by loading.

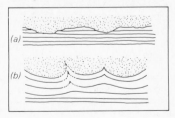

71

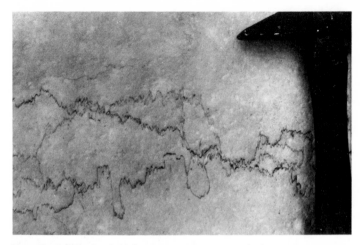

**Fig. 5.35** Stylolites in micritic limestone. Cretaceous, E. England.

mon in limestones, and calcite and dolomite nodules sometimes of immense size (metres across) occur in sandstones. Nodules may be randomly disposed or concentrated along particular horizons.

Nodule shape can vary considerably from spherical to flattened, to elongate, to highly irregular. Some nodules are nucleated around fossils and others form within animal burrow systems, but the majority are not related to any pre-existing inhomogeneity in the sediment. Occasionally, nodules possess radial and concentric cracks, infilled with mineral matter. The cracks in these *septarian* nodules, usually of calcite or siderite, form through contraction (dewatering) of the nodules soon after their formation.

Nodules can form at various times during diagenesis; the majority of nodules in muddy sediments form during early diagenesis, before the main phase of compaction. Calcareous nodules are common in marine mudrocks generally (but they do form in soils as calcretes, see below); pyritic nodules are more common in organic-rich muddy marine sediments, while siderite is more common in organic-rich non-marine sediments. As a result of the early diagenetic origin, lamination in the host sediment is deflected around the nodules (Fig. 5.37), and the nodules can preserve the original lamina thickness. The amount of compaction that has taken place can be deduced from these nodules (Fig. 5.37). Fossils and burrows in early diagenetic nodules are protected from compaction and so are unbroken and uncompressed. Early diagenetic nodules forming close to the sediment–water interface can be exposed on the seafloor or reworked and thus encrusted and bored by organisms (the nodules act as hardgrounds, Section 5.4.2). Chert nodules in limestones and carbonate nodules in sandstones are also chiefly

**Fig. 5.36** Diagenetic calcareous nodules in mudrock. Devonian, Germany.

of early diagenetic origin. Late diagenetic nodules in muddy sediments form after compaction so that lamination is undeflected from host sediment through the nodules (Fig. 5.37).

Calcareous nodules (*calcrete* or *caliche*) develop in soils of semi-arid environments where evaporation exceeds precipitation. In the geological record they are typically found in red-bed sequences, in fluviatile floodplain siltstones especially, but they can also occur in marine sediments, formed when the latter became emergent. In the field, calcretes usually occur as pale-coloured micritic nodules, a few centimetres in diameter, often with a downward elongation (Fig. 5.38). They vary from randomly-scattered to densely-packed nodules; laminated and pisolitic textures can be present. The presence of calcretes is best con-

firmed by petrographic study, but a feature which can be clearly observed in the field when they are developed in coarser sediments, is the splitting of pebbles and grains. Calcretes are a useful palaeoclimatic indicator, but

**Fig. 5.37** Pre-compactional (early diagenetic) and post-compactional (late diagenetic) nodules in mudrocks. An estimate of the amount of compaction can be obtained from early diagenetic nodules as $[(a - b) \div a] \times 100\%$.

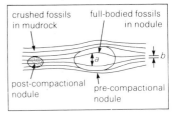

**Fig. 5.38** Calcrete (calcareous palaeosoil) consisting of elongate nodules of fine-grained calcite. Devonian, W. England.

they also reflect lengthy periods of non-deposition (tens of thousands of years), during which the pedogenic processes operated.

A scheme for describing nodules is given in Table 5.4.

## 5.6 Biogenic sedimentary structures

Many structures can be formed in sediments through the activities of animals and plants, in addition to the cryptalgal laminites and stromatolites described in Section 5.4.4. Structures produced vary considerably from poorly-defined and vague disruptions of lamination and bedding, to discrete and well-organized trace fossils (ichnofossils) which can be given a specific name. Trace fossils can often be interpreted in terms of the animal's activity which gave rise to the structure, but the nature of the animal itself is often difficult or impossible to deduce since different organisms may often have a similar mode of life. Also, an animal can produce different structures depending on its behaviour and on the sediment characteristics (grain-size, water content, etc). Burrow structures are commonly made by crustaceans, annelids, bivalves and echinoids, and surface trails and tracks by crustaceans, trilobites, annelids, gastropods and vertebrates.

Structures somewhat similar to burrows can be produced by the roots of plants, although the latter often contain carbonized cores (Section 5.6.4). Care should be taken not to mistake modern organic markings, such as polychaete and sponge borings and limpet tracks, frequently seen on rocks in the intertidal zone, for trace fossils. Confusion can also arise with some sedimentary structures, such as synaeresis cracks, tool marks, dewatering structures and diagenetic nodules, and spots and lineations in low-grade metamorphosed slates.

A scheme for the examination of trace fossils is given in Table 5.5.

## 5.6.1 Bioturbation

Bioturbation refers to the disruption of sediment by organisms. It varies from scattered burrow structures (often infilled with sediment of a different colour, composition or grain-size) to completely disrupted sediment which has a churned appearance and a loss of depositional sedimentary structures. A bioturbated bed can take on a nodular texture and segregations of coarser and finer sediment can occur.

## 5.6.2 Trace fossils

Trace fossils which are well developed and preserved are best considered in terms of their mode of formation; the principal groups are crawling, grazing and resting trace fossils, occurring on bedding surfaces and undersurfaces, and feeding and dwelling structures, mainly occurring within beds (Table 5.6, Fig. 5.39).

*Crawling traces* are produced by

**Table 5.4** Nodules: method of approach

| | |
|---|---|
| 1 | Determine composition of nodules and nature of host sediment. |
| 2 | Measure nodule size and spacing. |
| 3 | Describe shape and texture. |
| 4 | Look for nucleus (such as fossil). |
| 5 | Try to determine when formed: |
| | (a) early diagenetic: contain full-bodied fossils, locally reworked; pre-compactional so host sediment lamination is deflected; |
| or | (b) late diagenetic: contain crushed fossils, post-compactional so do not affect host-sediment lamination. |
| 6 | Check if calcrete (pedogenic nodules). |

animals in transit (not necessarily crawling), and so are usually straight or sinuous trails on bedding surfaces, contrasting with the more involved feeding and grazing structures (see below). They can be produced by many types of animal in any environment. Common crawling traces are made by crustaceans, trilobites and annelids. Vertebrates such as dinosaurs leave footprints as trace fossils.

*Resting traces* are made by animals resting on a sediment surface and leaving an impression of their body. Although a rare type of fossil, resting traces of starfish do occur.

*Grazing traces* occur on the sediment surface and are produced by deposit-feeding organisms systematically working the sediment for food. These biogenic structures usually consist of well-organized, coiled,

**Table 5.5** Trace fossils: how to describe them and what to look for

| | |
|---|---|
| 1   Sketch (or photograph) the structures; measure size, width, diameter, etc. Points to note for trails and tracks and burrows are:<br><br>2   *Trails and tracks* (on bed surfaces)<br>    (a) Examine trail: note whether regular or irregular pattern, whether trail is straight, sinuous, curved, coiled, meandering or radial.<br>    (b) Examine trail itself: if a continuous ridge or furrow note whether central division and any ornamentation (such as chevron pattern) exist: with appendage marks or footprints, measure size and spacing (gait) of impressions, look for tail marks.<br><br>3   *Burrows* (best seen within beds, also on bedding surfaces)<br>    (a) Describe shape and orientation to bedding; possibilities: horizontal, sub- | vertical, vertical; simple straight tube; simple curved or irregularly-disposed tube, U-tube. If branching burrow, note if regular or irregular branching pattern and any changes in burrow diameter.<br>    (b) Examine burrow wall: is the burrow lined with mud or pellets (look for scratch marks): are laminae in adjacent sediment deflected by the burrow?<br>    (c) Examine burrow fill: is it different from adjacent sediment (coarser or finer; richer or poorer in skeletal debris), is the fill pelleted; are there curved back-fill laminae within the burrow fill sediments?<br>    (d) Look for Spreite: laminae associated with U-shaped burrows. |

meandering and radiating patterns (e.g. Fig. 5.40). Grazing traces tend to occur in relatively quiet depositional environments.

*Dwelling structures* (Figs. 5.39, 5.41 and 5.42) are burrows varying from simple vertical tubes to U-shaped burrows, oriented either vertically, subvertically or horizontally to the bedding. With U-shaped burrows, concave-up laminae, referred to as Spreite, occur between and below the U-tube and form through the upward or downward movement of the animal in response to sedimentation or erosion. Where some burrowing organisms are rapidly buried, they can move up to regain their position relative to the sediment–water inter-

face; in so doing they leave behind a characteristic *escape structure* which deflects adjacent laminae to give a chevron structure. Other burrows, particularly those of crustaceans, are simple to irregularly branching systems, with walls composed of pellets or clay (Fig. 5.43). Some dwelling burrows possess curved laminae indicating back-filling by the animal.

*Feeding structures* are trace fossils developed within the sediment by deposit-feeding organisms searching for food. One of the commonest is a simple, non-branching, back-filled, horizontal to subhorizontal burrow (diameter 5 to 20 mm). Other feeding burrows are highly organized, in some cases with regular branching

**Table 5.6** Main features of trace fossil groups

> *Crawling traces:* trails, uncomplicated pattern; linear or sinuous.
>
> *Grazing traces:* more complicated surface trails, often symmetrical or ordered pattern, coiled, radial, meandering.
>
> *Resting traces:* impression of where animal rested during life (but not a fossil mould).
>
> *Dwelling structures:* simple to complex burrow systems but without suggestion of systematic working of sediment; burrows can be lined or pelleted.
>
> *Feeding structures:* simple to complex burrow systems often with well organized and defined branching pattern indicating systematic reworking of sediment.

patterns, in others with regular changes of direction, produced by the organism systematically working the sediment. Examples of feeding structures are shown in Figs. 5.39, 5.44 and 5.45.

## 5.6.3 Use of trace fossils in sedimentary studies

Biogenic sedimentary structures can give vital information for environmental interpretation, in terms of water depth, salinity, energy level etc., and they are especially valuable where body fossils are absent. It may be possible to recognize different suites of trace fossils in a sedimentary sequence so that various *ichnofacies* can be distinguished. Trace fossils can give an indication of sedimentation rate: intensely bioturbated horizons, beds with well-preserved, complex feeding and grazing traces, and bored surfaces (hardgrounds, Section 5.4.2) reflect slow rates of sedimentation; U-shaped burrows with Spreite and escape structures reflect rapid sedimentation.

**Fig. 5.39** Sketches of common dwelling and feeding burrows.

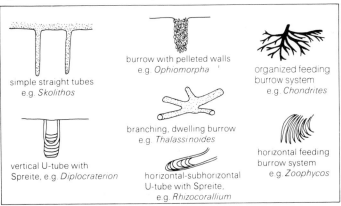

simple straight tubes
e.g. *Skolithos*

burrow with pelleted walls
e.g. *Ophiomorpha*

organized feeding burrow system
e.g. *Chondrites*

branching, dwelling burrow
e.g. *Thalassinoides*

vertical U-tube with Spreite, e.g. *Diplocraterion*

horizontal-subhorizontal U-tube with Spreite, e.g. *Rhizocorallium*

horizontal feeding burrow system
e.g. *Zoophycos*

**Fig. 5.40** *Nereites*, a grazing trace. Silurian, Wales.

Trace fossils can give an indication of sediment consistency; if the sediment is soft, then tracks made on a lamina can be transmitted through to underlying laminae (Fig. 5.46). Where burrows are either infilled with coarser sediment or are preferentially lithified during early diagenesis, surrounding sediments are frequently compacted around the burrow fills.

Trace fossils can show a preferred orientation, reflecting contemporaneous currents (bivalve resting traces commonly show this for example). Some can be used to show the way-up of strata.

### 5.6.4 Rootlet beds

Root systems of plants disrupt the internal structures of beds in a similar way to burrowing animals. Many roots and rootlets are vertically arranged while others are horizontally disposed; many branch freely. Roots are often carbonized, appearing as black streaks; many are preserved as impressions. The identification of rootlet beds indicates *in situ* growth of plants and thus subaerial conditions: coal seams may occur above rootlet beds. Plant debris is easily transported, however, so that plant-rich sediments are common. Examine a plant bed for rootlets; if it

**Fig. 5.41** *Skolithos*, a simple vertical dwelling burrow. Carboniferous, Wales.

**Fig. 5.42** *Rhizocorallium*, horizontal U-tube with spreite. Jurassic, N.E. England.

**Fig. 5.43** Poorly-organized burrow system, probably of a crustacean. Jurassic, S. England.

is just a collection of plant debris washed in, much of the material will be of the subaerial parts of the plant: leaves, stem and branches.

## 5.7 The geometry of sedimentary deposits and lateral facies changes

Some sedimentary rock units can be traced over large areas and show little change in character (that is facies) or thickness; others are laterally imper-

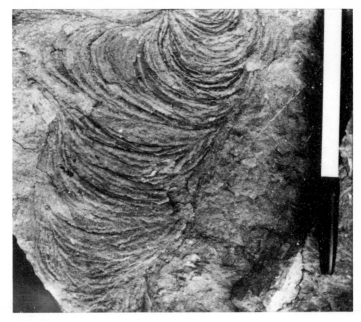

**Fig. 5.44** *Zoophycos*, a feeding burrow system. Carboniferous, N.E. England. Field of view 20 cm.

sistent. The geometry of sedimentary deposits should be considered on the scale of the individual bed or rock unit as seen in an exposure and on a larger, more regional, scale, in terms of the shape of the sediment body, or packet of a particular lithofacies, or group of related lithofacies.

The geometry of an individual bed or rock unit can be described as being *tabular* if laterally extensive, *wedge-shaped* if impersistent but with planar bounding surfaces, and *lenticular* if one or both of the bounding surfaces curve (Fig. 5.47). For the largerscale the terms *sheet* or *blanket* apply if the length to width ratio of the shape of the sediment body is around 1:1 and the sediment body covers a few to thousands of square kilometres. Elongate sediment bodies, where the length greatly exceeds the width, can be described as *ribbon* or *shoestring* if unbranching, *dendroid* if branching, and a *belt* if composite. Many elongate sand bodies are channel fills, oriented down the palaeoslope. Elongate sediment bodies can also form parallel to a shoreline, as in beach and barrier island developments. Sediment masses can be discrete entities, forming *pods* or *patches*, the latter term particularly applicable to some reef limestones. Coarse clastic sediments often form *fans* or *cones* where deposited at the toe of a slope. Examples include alluvial fan, fan-delta and submarine fan deposits.

**Fig. 5.45** *Chondrites*, a feeding burrow system. Devonian, S.W. England.

Facies, defined by the lithological, textural, structural and palaeontological features of the sedimentary rock, frequently change laterally as well as vertically in a sedimentary succession. This can involve a change in one or all of the parameters defining the facies. Lateral changes can be very rapid, over several or tens of metres, or more gradational, when the change takes place over several kilometres.

**Fig. 5.46** Trace made by animal on sediment surface transferred to lower laminae to form undertrace.

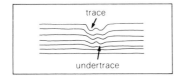

**Fig. 5.47** Common geometries of beds or rock units (on a scale of metres to tens of metres) and sediment bodies (on kilometre or regional scale).

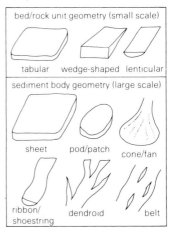

Facies changes reflect changes in the environmental conditions of sedimentation (Chapter 8).

Observations of the geometry of individual beds and rock units normally present no problem. In a quarry or cliff exposure, follow beds laterally to check on their shape and then make notes and sketches (or take photographs). With the larger-scale geometry of sediment bodies, if exposures are very good as in some mountainous and vegetation-free regions, it may be possible to see the lateral changes and sediment body shape directly. Where exposure is limited, make detailed logs across the same part of the rock sequence at several or many localities. To be sure that sections are equivalent it is necessary to have either a laterally continuous horizon in the sequence, or the presence of the same zonal fossils. If lateral facies changes are suspected in an area of poor exposure then detailed mapping and logging of all available exposures may be required to demonstrate such changes.

# 6

# *Fossils in the field*

## 6.1 Introduction

Fossils are an important component of sedimentary rocks. First and foremost, they can be used for stratigraphic purposes, to determine the age of the rock sequence and to correlate it with sequences elsewhere. The identification of fossils to the species level is not easy and in most cases is best left to the specialist. Fossils are of great use in the environmental interpretation of sedimentary rocks and in this context many useful observations can be made in the field by the non-specialist with a keen eye. Fossils can tell you about the depth, level of turbulence, salinity and sedimentation rate; they can record palaeocurrent directions and give information on palaeoclimatology. In some cases the whole environmental interpretation of a sedimentary sequence may depend on the presence of just a few fossils and occasionally an interpretation has been shattered by the discovery of new fossils. Field observations on fossils should consider their distribution, preservation and relation to the sediment, their associations and diversity, and their diagenesis. A checklist for the examination of fossils in the field is given in Table 6.1

## 6.2 Fossil distribution and preservation

Pay attention to the distribution of the fossils while studying or logging a sequence of sediments. Fossils can be evenly distributed throughout a rock unit without any preferential concentration in a certain horizon or particular bed: this generally only occurs where the sediment is homogeneous throughout. Often fossils are not evenly distributed but occur preferentially in certain beds, lenses, pockets or in build-ups (a reef or shell bank for example). Always examine the types of fossil present and determine their relative distribution. See if there is a correlation between fossil type and lithofacies. Concentrations of fossils at particular levels can arise from current activity or preferential growth through favourable environmental conditions.

### 6.2.1 In situ fossil accumulations

Where favourable conditions existed it is likely that fossils will be preserved intact, with little breakage or disarticulation, and that at least some of the organisms which lived on or in the sediment will be in their growth position. Fossils which are commonly

**Table 6.1** Checklist for the examination of fossils in the field

## A  Distribution of fossils in sediment

1 *Fossils largely in growth position*

(a) Do they constitute a reef?—characterized by: colonial organisms; interaction between organisms (such as encrusting growth); presence of original cavities (infilled with sediment and/or cement) and massive, unbedded appearance (Section 3.5.3):
  (i) describe growth forms of colonial organisms; do these change up through reef?
  (ii) are some skeletons providing a framework?

(b) If non-reef, are fossils epifaunal or infaunal? if epifaunal how have fossils been preserved: by smoothering, for example?

(c) Do epifaunal fossils have a preferred orientation, reflecting contemporary currents? if so, measure.

(d) Are fossils encrusting substrate, i.e., is it a hardground surface?

(e) Are the plant remains rootlets?

2 *Fossils not in growth position*

(a) Are they concentrated into pockets, lenses or laterally-persistent beds or are they evenly distributed throughout the sediment?

(b) Do fossils occur in a particular lithofacies? are there differences in the faunal content of different lithofacies?

(c) If fossil concentrations occur, what proportion of fossils are broken and disarticulated? Are delicate skeletal structures preserved, such as spines on shells? Check sorting of fossils, degree of rounding; look for imbrication, graded bedding, cross-bedding, scoured bases and sole structures.

(d) Do fossils show a preferred orientation? if so, measure.

(e) Have fossils been bored or encrusted?

(f) Note degree of bioturbation and any trace fossils present.

## B Fossil assemblages and diversity

1 Determine the composition of the fossil assemblages by estimating the relative abundance of the different fossil groups in a bed or on a bedding plane.

2 Is the fossil assemblage identical in all beds of the section or are there several different assemblages present? if the latter, do the assemblages correlate with different lithofacies?

3 Consider the degree of reworking and transportation; does the fossil assemblage reflect the community of organisms which lived in that area?

4 Consider the composition of the fossil assemblage; for example, is it dominated by only a few species, are they euryhaline or stenohaline? are certain fossil groups conspicuous by their absence? do all fossil groups present have a similar mode of life? do pelagic forms dominate? are infaunal organisms absent?

## C Diagenesis of fossil skeletons

1 Is original mineralogy preserved or have skeletons been replaced: dolomitized, silicified, hematized, pyritized, etc?

2 Have fossils been dissolved out to leave moulds?

3 Do fossils occur preferentially in nodules?

4 Are fossils full-bodied or have they been compacted?

found in growth position include brachiopods (Fig. 6.1), some bivalves, corals (Fig. 6.2), bryozoans and stromatoporoids. It is worth noting the general level of infaunal activity (amount of bioturbation) in conjunction with the body fossils present.

Fossils can be organized into reefs or build-ups (Section 3.5.3); in these the majority of the fossils are in growth position, perhaps with some organisms growing over and upon each other. Colonial organisms can dominate and the rock characteristically has a massive, unbedded appearance. Cavities, both large and small, perhaps infilled with internal sediment and calcite cement, are common in reefal limestones.

## 6.2.2  Current accumulations

Concentrations of skeletal material (Fig. 6.3) arising from currents can form in a number of ways. Transportation of skeletal debris by storm currents leads to the deposition of *storm beds*. These tend to be laterally-persistent, possessing sharp often scoured bases. They can either show a normal size grading, good sorting of particles and lamination, or little sorting, the deposit having a 'dumped' appearance with a wide range of grain-sizes and no internal structure. Storm beds vary from a few cm to a few tens of centimetres in thickness and are characteristic of subtidal shelf and platform sequences. Fossil accumulations can also form through the winnowing action of weaker currents which remove finer sediment and skeletal grains. Such *fossil lag deposits* are usually impersistent lenses and pockets; they too occur in shallow shelf carbonate sequences. Fossil-rich beds can also be formed through reworking of sediment by migrating tidal channels.

The degree of current activity and reworking affects the proportion of broken and disarticulated carbonate skeletons within a fossil concentrate.

**Fig. 6.1**  Brachiopods (productids) in growth position. Carboniferous, Wales.

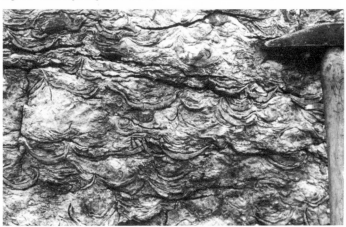

**Fig 6.2** Colonial corals (*Lithostrotion* sp.) in growth position. Carboniferous, Wales.

With an increasing level of agitation, fossils grade from perfect preservation, with all delicate structures intact and adjoined, to poor preservation where fossils are abraded and broken. Points to look for with specific fossils are *crinoids:* the length of the stems and whether all ossicles are separated, if the calyx is present whether this is attached to the stem or not; *bivalves, brachiopods* and *ostracods:* whether the valves are disarticulated or articulated, if the former whether there are equal numbers of each valve, if the latter whether the valves are open or closed; some *brachiopods* and *bivalves:* whether the spines for anchorage are still attached; *trilobites:* whether the exoskeleton is whole or incomplete.

### 6.2.3 Preferred orientations

Elongate shells or skeletons affected by currents frequently have a preferred orientation to their long-axes (Fig. 6.4). This alignment can be parallel to or less commonly normal to the current if the skeletal fragments were subject to rolling. Preferred orientations are commonly found with crinoid stems, graptolites, cricoconarids, elongate bivalve shells, solitary corals, belemnites, orthoconic nautiloids and plant fragments. Some orientations may reflect current directions during life. Chapter 7 deals with palaeocurrent analysis.

## 6.3 Fossil associations and diversity

### 6.3.1 Fossil assemblages

Those present, and their relationship to each other, give useful environmental information. First determine the fossil assemblage qualitatively by estimating the relative abundance of

**Fig. 6.3** Current accumulation of brachiopod shells. Many are still articulated and some possess geopetal structures. Carboniferous, N.W. England. Field of view 30cm.

the different fossil groups. For a precise analysis of the assemblage large blocks of the sediment need to be broken up carefully and all species identified and counted. This is best undertaken in the laboratory. If good bedding plane exposures are available count the number of each fossil species in a quadrat; a square of a particular area, one square metre, is usually taken. By carefully analyzing the fossil assemblage from different beds in a section or different but coeval lithofacies over an area, changes in the assemblage can be recognized. Assemblages can be described by their dominant members; choose one or several characteristic forms present: e.g. Productid-*Lithostrotion* assemblage (common in the Lower Carboniferous), *Micraster*-terebratulid-sponge assemblage (common in the Upper Cretaceous).

A fossil assemblage is also a death assemblage. Many such assemblages are composed of the remains of animals which did not live in the same area. The skeletal debris was brought together by currents and so generally consists of broken and disarticulated skeletons. Some death assemblages consist of the skeletons of organisms which did live in the same general area. In these cases, some fossils may occur in their original life position and skeletal transportation has been minimal. Reefs and other build-ups are obvious examples of *in situ* death assemblages.

Where little transport of skeletal material has taken place after death, the fossil assemblage will reflect the *community* of organisms which lived in that area. A community can be referred to either by a dominant species, in the same way as an assem-

**Fig. 6.4** Graptolites showing a preferred orientation. Ordovician, Wales.

blage, or by reference to the lithofacies (e.g., a muddy sand community). The ascribing of an assemblage to a community is an important step, since a community is dependent on environmental factors and changes in community can indicate changes in the environment. Once a community has been recognized, it is possible to look at the species present and deduce the roles various organisms played in that community. It must be remembered that much, if not most, of the fossil record is not preserved. There is obviously a bias towards the preservation of animal hard parts. In studying a community, thought should be given to animals and plants for which there is only circumstantial or no evidence. Trace fossils, pellets, coprolites and algal laminations are important in this respect.

One feature to look for in an assemblage is the presence of encrusting and boring organisms. Large skeletal fragments can act as substrates for others; oysters, bryozoans, barnacles, certain inarticulate brachiopods, algae and serpulid worms frequently encrust other skeletons. Boring organisms such as serpulids, lithophagid bivalves and sponges can attack skeletal fragments and other hard substrates (such as hardground surfaces, Section 5.4.2), producing characteristic holes and tubes. Boring and encrusting of skeletal debris tends to be more common where sedimentation rates are low. Burrows in sediment are also more common in such situations (Section 5.6)

## 6.3.2 Species diversity

The number and type of species present in an assemblage depend on environmental factors. Where these factors (depth, salinity, agitation, substrate, oxygenation etc.) are at an optimum, there is maximum species diversity. Infaunal and epifaunal benthic, nektonic and planktonic organisms are all present. Where there are environmental pressures, species diversity is lowered and certain aspects of the fauna and flora may be missing. However, in these situations, species that are present and can tolerate the environment may occur in great numbers. With increasing depth pelagic fossils such as fish, graptolites, cephalopods, posidonid bivalves and some ostracods for example, will dominate. The same situation is found with an increasing degree of stagnation: benthic organisms will eventually be excluded altogether, so that only pelagic organisms are present. With higher or lower salinities than normal marine, many species are excluded completely. Groups tolerant of normal marine conditions only (stenohaline forms) are corals, bryozoans, stromatoporoids and trilobites; many specific genera and species of other groups are also stenohaline. Some fossil groups (euryhaline forms) are able to tolerate extremes of salinity; certain bivalves, gastropods, ostracods and charophyte algae are examples. Where sediments contain large numbers of only a few species of such groups, then hypersaline or hyposaline conditions should be suspected. In some cases, organism skeletons show a change in shape or size with extremes of salinity. Hypersaline conditions may well be indicated by the presence of evaporite pseudomorphs (Section 3.6).

## 6.4 Skeletal diagenesis

The original composition of fossil skeletons is frequently altered during diagenesis. Many carbonate skeletons are and were composed of aragonite when the animal was living. With the majority of such fossils the aragonite has been replaced by calcite. Other minerals which replace fossils include dolomite, pyrite, hematite and silica. Occasionally fossils can be dissolved out completely so that only moulds are left; this can happen preferentially to fossils originally composed of aragonite.

Where early diagenetic nodules occur in a mudrock sequence fossils present within the nodules will be better preserved, that is less compacted, compared with fossils in the surrounding mudrock. Nodule nucleation can take place preferentially around fossils; for example the decay of fish in sediment can set up a chemical microenvironment conducive for mineral precipitation.

# 7

# *Palaeocurrent analysis*

## 7.1 Introduction

Palaeocurrents often reflect the regional palaeoslope, hence their analysis is a vital part of sedimentary rock studies, providing data on the palaeogeography and useful information for facies interpretation. The measurement of palaeocurrents in the field should become a routine procedure; a palaeocurrent direction is an important attribute of a lithofacies and necessary for its complete description.

Many different featues of a sedimentary rock can be used as palaeocurrent indicators. Some structures record the direction of movement (azimuth) of the current while others only record the line of movement (trend). Of the sedimentary structures, the most useful are crossbedding and sole structures (flute and groove casts); but other structures also give reliable results.

## 7.2 Palaeocurrent measurements

The more measurements you can take from either a bed or a series of beds, the more accurate is the palaeocurrent direction you obtain, although an important consideration is the variability (spread) of the measurements.

First, assess the outcrop. If only one lithofacies is present measurements can be collected from any or all of the beds. If measurements from one bed (or many beds at an outcrop) are all similar (a unimodal palaeocurrent pattern, Fig. 7.1) there is no value in taking a large number of readings. Some 20 to 30 measurements from an exposure would be sufficient to give an accurate vector mean (Section 7.4). You should then find other outcrops of the same lithofacies in the immediate vicinity and farther afield, so that the palaeocurrent pattern over the area can be deduced. If within a bed readings vary considerably you will need to collect a large number (more than 20 or more than 50 depending on the variability) to ascertain the mean direction.

Measurements collected from different sedimentary structures should be kept apart, at least initially. If they are very similar they can be combined. Also keep separate measurements from different lithofacies at an exposure; they may have been deposited by different types of current, or currents from different directions. Readings should be tabulated in your field notebook.

The measurements you have taken may not represent the current direction if either the shape or the orientation (or both) of the sedimentary

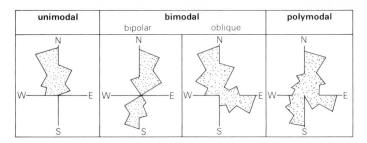

**Fig. 7.1** The four types of palaeocurrent pattern, plotted as rose diagrams (with 30° intervals). The convention is to plot palaeocurrent azimuthal data in a 'current to' sense, so that in the unimodal case above, the current was flowing from the south towards the north.

structures has been changed by tectonism. It is important to appreciate that two changes can occur: tilt and deformation. A simple change in the inclination of the plane, of which the sedimentary structure is a part, is described as the tilt. Tilt does not change the shape of a sedimentary structure. Processes which change the shape of a sedimentary structure are described as deformation.

To ascertain the direction of palaeocurrents from structures which have been tilted it is necessary to remove the effects of tilting, a simple process that is described later. To do the same with deformed sedimentary structures is not simple and requires an accurate assessment of the strain the rock mass containing the sedimentary structure has undergone: a description of how this is done is beyond the scope of this handbook but an excellent account can be found in Ragan (1973). The tell-tale signs of deformed rock masses are the presence of such features as cleavage, minor folding, metamorphic fabrics and deformed fossils.

The following steps should be taken to correct the azimuth of a linear structure that has been changed by tilt.

1 Measure the angle and direction of dip of the bedding surface (or undersurface) containing the sedimentary structure and plot the surface as a great circle on a stereonet.
2 Mark on the great circle the acute angle between the direction of the sedimentary structure and the strike of the bedding surface (i.e. the pitch, or rake, of the sedimentary structure).
3 Rotate the great circle to the horizontal and with it the pitch of the sedimentary structure: the effect of tilt on the azimuth of the structure is now removed and the azimuth can be calculated (Fig. 7.2).

The following steps should be taken to correct the orientation of a planar structure (e.g. cross-bedding) that has been changed by tilt.

1 Measure the dip and direction of dip of the planar sedimentary structure and plot the pole for this plane on a stereonet.
2 Similarly plot the pole of the bed that contains the structure.
3 Rotate the latter to the horizontal and rotate the former by the same number of degrees.
4 The new position of the pole to

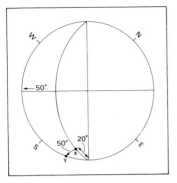

**Fig. 7.2** An example of correcting the orientation of a linear structure for tectonic tilt. A bed dips at 50° to 225° and has a linear structure with a pitch of 20° to the SE. The bed is plotted as a great circle on a stereonet and the pitch of the structure is marked on that circle (X). The bed is restored to the horizontal and with it the structure along a small circle to give its original orientation at point Y. On rotating back, the azimuth of Y is given: 155°.

the planar sedimentary structure now gives its orientation with tilt removed (Fig. 7.3).

Finally a note of warning: if tilted strata are part of a fold whose axis of folding is inclined (i.e. plunging) the fold axis must be brought to horizontal, so changing the orientation of tilt, before the tilt is removed. Having completed these corrections can you be certain that the palaeocurrent direction is that at its time of formation? Unfortunately not, for rotations of outcrop that have occurred about a vertical axis should be known and removed; rarely are they known.

A stereonet for making corrections for tectonic tilt is given on the inside back cover.

## 7.3 Structures for palaeo-current measurement

### 7.3.1 Cross-bedding

This is one of the best structures to use but first determine what type of cross-bedding is present (Section 5.3.3). If it has formed by the migration of subaqueous dunes and sand waves (much is of this type) or aeolian dunes, then it is eminently suitable for palaeocurrent (or palaeowind) measurement. Check whether the cross-bedding is of the planar (tabular) or trough type (Figs. 5.12 and 5.3.3b).

With *planar cross-bedding*, the palaeocurrent direction is simply given by the direction of maximum angle of dip. If the exposure is three-dimensional, or two-dimensional with a bedding plane surface there is no problem in measuring this directly. If there is only one vertical face showing the cross-bedding then taking readings is less satisfactory since it is just the orientation of the face that is being measured, which is unlikely to be exactly in the palaeo-current direction. Close scrutiny of the rock face may enable you to see a little of the cross-bed surface and so determine the actual dip direction. If even this is impossible there is no alternative but to measure the orientation of the rock surface.

With *trough cross-bedding*, it is essential to have a three-dimensional exposure or one with a bedding plane section, so that the shape of the cross strata is clearly visible and the dip direction down the trough axis can be measured accurately. Because of the shape of the cross-beds, vertical sections can show cross-bedding dipping 90° to current direction. Vertical sections of trough cross-

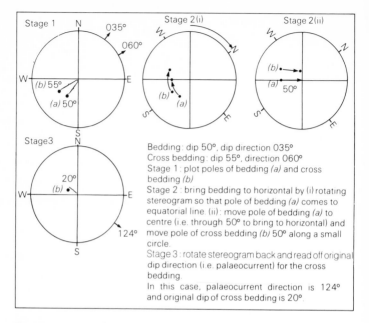

Figure content:

Stage 1

035°
060°

(b) 55°
(a) 50°

Stage 2(i)

(b)
(a)

Stage 2(ii)

(b)
(a) 50°

Stage 3

20°
(b)

124°

Bedding: dip 50°, dip direction 035°
Cross bedding: dip 55°, direction 060°
Stage 1: plot poles of bedding (a) and cross bedding (b)
Stage 2: bring bedding to horizontal by (i) rotating stereogram so that pole of bedding (a) comes to equatorial line. (ii): move pole of bedding (a) to centre (i.e. through 50° to bring to horizontal) and move pole of cross bedding (b) 50° along a small circle.
Stage 3: rotate stereogram back and read off original dip direction (i.e. palaeocurrent) for the cross bedding.
In this case, palaeocurrent direction is 124° and original dip of cross bedding is 20°.

**Fig. 7.3** Correction of cross-bedding for tectonic tilt using stereographic projection.

bedding are thus unreliable for palaeocurrent measurements and should only be used as a last resort.

### 7.3.2 Ripples and cross-lamination

Palaeocurrent directions are easily taken from current ripples and cross-lamination. The asymmetry of the ripples (steeper, lee-side, downstream) and the direction of dip of the cross-laminae are easily measured. However, ripples and the cross-lamination they give rise to, are often produced by local flow directions which do not reflect the regional palaeoslope. In turbidite beds, for example, cross-lamination within the bed may vary

considerably in orientation and differ substantially from the palaeocurrent direction recorded in the sole structures. The cross-lamination forms when the tubidity current has slowed down somewhat and is wandering or meandering across the seafloor. In spite of their shortcomings, if there is no other more suitable directional structure present (cross-bedding or sole structures) it is always worth recording the orientation of the ripples and cross-lamination.

Wave-formed ripples are small-scale structures which record local shoreline trends and wind directions; their crest orientation should be measured, or if present the direction of dip of internal cross-lamination.

### 7.3.3 Sole structures

Flute casts provide a current azimuth (Section 5.2.1); some tool marks will also provide a current azimuth, if not at least a line of movement; groove casts provide a line of movement. Flute casts are generally all oriented in the same direction and so several measurements from one bed together with measurements from other beds at the exposure are sufficient. With groove casts, there can be a substantial variation in orientation and a larger number (more than 20) should be measured, from which the vector mean can be calculated for each bed (Section 7.4).

### 7.3.4 Preferred orientations of clasts and fossils

Pebbles and fossils with an elongation ratio of at least 3:1 can be aligned parallel to or normal to prevailing currents. Check for such a preferred orientation and if you either suspect or see its presence, measure and plot the elongation of a sufficient number of the objects. In many cases you will obtain a bimodal distribution with one mode, that parallel to the current, dominant. Pebbles, grains and fossils mostly give a line or trend of movement; with some fossils, orthoconic cephalopods and high-spired gastropods for example, a direction of movement can be obtained (the pointed ends preferentially-directed upstream).

### 7.3.5 Other directional structures

Parting lineation records the trend of the current and presents no problem in its measurement. Channels and scour structures can also preserve the trend of the palaeocurrent. Slump folds record the direction of the palaeoslope down which slumping occurred (fold axes parallel to strike of palaeoslope, anticlinal overturning downslope). Glacial striations on bedrock show the direction of ice-movement.

## 7.4 Presentation of results and calculation of vector means

Palaeocurrent measurements are grouped into classes of 10°, 15°, 20° or 30° intervals (depending on number of readings and variability) and then plotted on a rose diagram, choosing a suitable scale along the radius for the number of readings. For data from structures giving current azimuths the rose diagram is conventionally constructed showing the current-to sense (in contrast to wind roses).

Although the dominant palaeocurrent (or palaeowind) direction will usually be obvious from a rose diagram, for accurate work it is necessary to calculate the mean palaeocurrent direction (that is the *vector mean*). It is also worth calculating dispersion (or variance) of the data. Vector means and dispersion can only be calculated for a unimodal palaeocurrent pattern (Fig. 7.1).

To deduce the vector means from structures giving azimuths each observation is considered to have both direction and magnitude (the magnitude is generally considered unity but it can be weighted); the north–south and east–west components of each vector are then calculated by multiplying the magnitude by the cosine and sine of the azimuth respectively. The components are summed and a division of the summed E–W components by the N–S components gives the tangent of the resultant vector; that is the vector mean.

E–W component $= \Sigma n \sin \sigma$

N–S component $= \Sigma n \cos \sigma$

$$\tan \bar{\sigma} = \frac{\Sigma n \sin \sigma}{\Sigma n \cos \sigma}$$

where $\sigma$ = azimuth of each observation from 0 to 360°

$n$ = observation vector magnitude, generally 1, but if data are grouped into classes (0–15, 16–30, 31–45, etc.) then it is the number of observations in each group

$\bar{\sigma}$ = azimuth of resultant vector (i.e. vector mean)

If trends only can be measured then each observation, measured in the range 0 to 180°, is doubled before the components are calculated:

E–W component $= \Sigma n \sin 2\sigma$

N–S component $= \Sigma n \cos 2\sigma$

$$\tan 2\bar{\sigma} = \frac{\Sigma n \sin 2\sigma}{\Sigma n \cos 2\sigma}$$

The magnitude $(r)$ of the vector mean gives an indication of the dispersion of the data, comparable to the standard deviation or variance of linear data:

$r = \sqrt{(\Sigma n \sin \sigma)^2 + (\Sigma n \cos \sigma)^2}$ for azimuthal data, and

$r = \sqrt{(\Sigma n \sin 2\sigma)^2 - (\Sigma n \cos 2\sigma)^2}$ for non-azimuthal data.

To calculate the magnitude of the vector mean in terms of a per cent $(L)$

$$L = \frac{r}{\Sigma n} \cdot 100$$

A vector magnitude of 100% means that all the observations have either the same azimuth or lie within the same azimuth group. In a vector magnitude of 0% the distribution is completely random. There would be no vector mean in this case.

For further discussion and methods for testing the significance of two-dimensional orientation distributions see Potter and Pettijohn (1977).

Vector means can easily be calculated in the evenings after fieldwork and then entered on the geological map or graphic sedimentary log. The raw data should be kept.

## 7.5 Interpretation of the palaeocurrent pattern

Four types of palaeocurrent pattern can be obtained (Fig. 7.1): *unimodal:* where there is one dominant current direction; *bimodal bipolar:* two opposite directions; *bimodal oblique:* two current directions at an angle less than 180°; and *polymodal* where there are several dominant directions. Analysis of the palaeocurrent pattern needs to be combined with a study of the lithofacies for maximum information. The features of the palaeocurrent (palaeowind) pattern of the principal depositional environments, fluvial, deltaic, aeolian sand, shoreline-shallow shelf and turbidite basin, are shown in Table 7.1. In fluvial, deltaic and turbidite-basin environments palaeocurrents are related to local or regional palaeoslopes, although in the case of turbidite-basins, currents also flow along basin axes. In shoreline-shelf environments and deserts, palaeocurrent and palaeowind directions are not related to a regional palaeoslope and interpretations of the palaeogeography need to be made with care.

**Table 7.1** Palaeocurrent patterns of principal depositional environments, together with *best* and other directional structures.

| environment | directional structures | typical dispersal patterns |
|---|---|---|
| aeolian | large scale cross bedding | unimodal common, also bimodal and polymodal; dependent on wind directions/dune type |
| fluvial | *cross bedding*, also parting lineation, ripples | unimodal down palaeoslope, dispersion reflects river sinuosity |
| deltaic | *cross bedding*, also parting lineation, ripples | unimodal directed offshore, but bimodal or polymodal if marine processes important |
| marine shelf | *cross bedding*, also ripples, fossil orientations | bimodal common through tidal current reversals but can be normal or parallel to shoreline; unimodal and polymodal patterns also |
| turbidite basin | *flutes*, also grooves, parting lineation, ripples | unimodal common, either downslope or along basin axis if turbidites, parallel to slope if contourites |

# 8

# What next? Facies identification and analysis

Having collected all the field data on the sedimentary sequence it remains to interpret the information. Many studies of sedimentary rocks are concerned with elucidating the conditions, environments and processes of deposition. The field data are particularly pertinent to such considerations. Other studies are concerned more with particular aspects of the rocks, such as the possibility of there being economic minerals and resources present, the origin of specific structures or the diagenetic history. Following the fieldwork, laboratory examination of the rocks is necessary in many instances, not least to deduce or confirm sediment composition.

## Facies analysis

If the aim of the study is to deduce the depositional processes and environments then, with all the field data at hand, the *facies* present within the sequence should be identified. A facies is defined by a particular set of sediment attributes: a characteristic lithology, texture, suite of sedimentary structures, fossil content, colour, geometry, palaeocurrent pattern, etc. A facies is produced by one or several processes operating in a depositional environment, although of course the appearance of the facies can be con-

siderably modified by post-depositional, diagenetic processes. Within a sedimentary sequence there may be many different facies present, but usually the number is not great. A particular facies is often repeated several or many times in a sequence. A facies may also change vertically or laterally into another facies by a change in one or several of its characteristic features. In some cases it will be possible to recognize subfacies, sediments which are similar to each other in many respects but which show some differences.

Facies are best referred to objectively in purely descriptive terms, using a few pertinent adjectives; examples could be cross-bedded, coarse sandstone facies or massive pebbly mudstone facies. Facies can be numbered or referred to by letter (facies A, facies B, etc.). However, facies are frequently referred to by their environment such as fluviatile facies or lagoonal facies, or by their depositional mechanism, such as turbidite sandstone facies or storm bed facies. In the field and during the early stages of the study, facies should be referred to only in the descriptive sense. Interpretations in terms of process and/or environment can come later.

After logging and examining a sequence in detail, look closely at the log with all the sediment attributes recorded and look for beds or units with similar features. Take the depositional structures first since these best reflect the depositional process; then check the texture, lithology and fossil content. You will probably find that there are a number of distinct sediment types with similar attributes; these will be of the same facies. Name or number them for reference.

Once the various facies have been differentiated, they can then be interpreted by reference to published accounts of modern sediments and ancient sedimentary facies, and by the erection of facies models. Many textbooks contain reviews of modern depositional environments, their sediments and their ancient analogues: see Further Reading, but particularly Reading (1978) and Walker (1979). Some facies are readily interpreted in terms of depositional environment and conditions, while others are not environmentally diagnostic and have to be taken in the context of adjacent facies. As an example, a fenestral pelleted limestone will almost certainly have been deposited in a tidal flat environment whereas a cross-bedded coarse sandstone could be fluvial, lacustrine, deltaic, shallow-marine, even deep-marine, and could have been deposited by a variety of different processes. A number of depositional processes produce distinctive facies but can operate in several environments; for example, turbidity currents occur in lacustrine and marine basins.

Facies interpretation is often facilitated by considering the vertical *facies sequence*. Where there is a conformable vertical succession of facies, with no major break, the facies are the products of environments which

were originally laterally adjacent. This concept has been appreciated since Johannes Walther expounded his *Law of Facies* in 1894. The vertical succession of facies is produced by the prograding (building out) or lateral migration of one environment over another. Where there are breaks in the sequence, seen as sharp or erosional contacts between facies, then the facies sequence need not reflect laterally-adjacent environments, but could well be the products of widely-separated environments. Other environments whose sediments have been eroded could be represented by the break.

In a sedimentary sequence it is often found that groups of facies occur together to form *facies associations*. The facies comprising an association are generally deposited in the same broad environment, in which there are either several different depositional processes operating or fluctuations in the depositional conditions.

The relationships between facies in a succession, or within a facies association, can be completely random. However, in many instances there is a regular repetition of the various facies to give *cycles* of sedimentation (also called cyclothems or rhythms). With many cyclic sequences there is a distinct boundary at the top (and therefore also bottom) of each cycle. This is frequently an emergence horizon (rootlet bed, soil horizon or coal seam) or an erosion surface. There are often systematic changes in the character of the sediments up through a cycle, such as a gradual increase or decrease in grain-size, change in lithology and fossil content, or scale and abundance of sedimentary structures. Cycles are typically well developed in fluviatile and deltaic clastic sediments and in the field such cycles

99

are often immediately apparent because of major lithological and textural changes within them. Cycles also occur in the sediments of many other environments. With limestone sequences the cycles may not be so obvious in the field, so that laboratory work to determine microfacies may be necessary.

Although with some sequences cycles may be obvious in the field, or on examination of the graphic log, in other cases a cyclic pattern of sedimentation may be more subjective or not apparent at all. Various techniques are available to test a sequence for facies relationships, randomness and repetitions. Simple methods involve counting the number of times each facies is overlain by the others and noting the type of boundary (sharp or gradational) between facies. A facies relationship diagram can be constructed from the data or the data can be tabulated (in a 'data array') and

compared with a table of values for a random arrangement of the facies: for details see Reading (1978) or Walker (1979). For a more rigorous statistical treatment of the data Markov chain analysis can be used.

The principal features of the main depositional facies are given in Tables 8.1 to 8.10. These are superficial indeed and are only intended to give a broad indication of the appearance of the various facies. They are in no way complete and in fact it is totally impossible to summarize adequately the features of the various facies in a page or two. Facies interpretation requires much more thought and industry than just looking at a few tables. Very detailed analyses of facies can now be made and for help you should refer in the first instance to the textbooks cited in the 'Further Reading' section and then to the scientific journals.

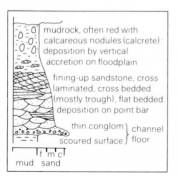

Fig. 8.1 Typical fining-up sequence produced by lateral migration of meandering stream. Such sequences vary from a few metres to a few tens of metres thick. Lateral accretion surfaces may occur within the sandstone member.

**Table 8.1**  General features of fluviatile facies

*Deposition:* is complex, alluvial systems include meandering streams with well-developed floodplains, braided streams, and alluvial fans. In the first, lateral migration of channels is characteristic, with overbank sedimentation and crevasse splays on floodplains. Channel processes dominate in braided streams, and on alluvial fans, stream and sheet floods and debris flows occur.

*Lithologies:* from conglomerates through sandstones to mudrocks; thin intraformational conglomerates common; many sandstones are lithic or arkosic.

*Textures:* many stream-deposited conglomerates have a pebble-support fabric with imbrication, debris flow conglomerates are matrix-supported; many fluviatile sandstones are red and consist of angular to rounded grains, with moderate sorting.

*Structures:* fluviatile sandstones show tabular and trough cross-bedding, flat bedding + parting lineation, channels and scoured surfaces; finer sandstones show ripples and cross lamination; stream-deposited conglomerates are often lenticular with crude cross-bedding; mudrocks are often massive, with rootlets and calcareous nodules (calcrete).

*Fossils:* plants dominate (fragments or *in situ*), fish bones and scales, freshwater molluscs.

*Palaeocurrents:* unidirectional, but dispersion depends on stream type.

*Geometry:* sand bodies vary from ribbons to belts to fans.

*Facies sequences:* depend on type of alluvial system: alluvial fan sequences may show an overall coarsening or fining up depending on climatic/tectonic changes; meandering streams produce fining-upward cross-bedded sandstone units up to several metres thick with lateral accretion surfaces, interbedded with mudrocks, often containing calcretes; sandy braided streams produce lenticular cross-bedded sandstones with few mudrock interbeds.

*Relevant sections:* 3.2, 3.3, 3.4, 5.2, 5.3

*Relevant figures:* 4.4, 5.3 to 5.13, 5.18, 5.38, 8.1

**Table 8.2** General features of aeolian facies

*Deposition:* wind-blown sand is typical of deserts but also occurs along marine shorelines.

*Lithology:* clean (matrix-free) quartz-rich sandstones, no mica.

*Textures:* well-sorted, well-rounded sand grains ('millet-seed'); possibly with a frosted (dull) appearance; sandstones often stained red through hematite; any pebbles may be wind-faceted.

*Structures:* dominantly large-scale cross-bedding (set heights several to several 10's of metres); cross-bed dips up to 35°.

*Fossils:* rare, occasional vertebrate footprints and bones.

*Facies associations:* water-lain sandstones and conglomerates may be associated; also playa-lake mudrocks and evaporites and arid-zone soils.

*Relevant sections:* 5.3.2c, 5.3.3i

*Relevant figure:* 5.17

---

**Table 8.3** General features of lacustrine facies

*Deposition:* in lakes which vary in size, shape, salinity and depth. Waves and storm currents important in shallow water, turbidity currents, often river underflows, in deeper water. Biochemical and chemical precipitation common. Strong climatic control on lake sedimentation.

*Lithologies:* diverse including conglomerates through sandstones to mudrocks, limestones, marls, evaporites, cherts, oil shales and coals.

*Structures:* wave-formed ripples, desiccation cracks and stromatolites common in lake shoreline sediments; rhythmic laminations, possibly with synaeresis cracks, typical of deeper water lake deposits, together with interbedded graded sandstones of turbidity current origin.

*Fossils:* non-marine invertebrates (especially bivalves and gastropods); vertebrates (footprints and bones); plants, especially algae.

*Facies sequences:* often reflect changes in water level through climatic or tectonic events.

*Facies associations:* fluviatile and aeolian sediments often associated; soil horizons may occur within lacustrine sequences.

*Relevant section:* 5.3

*Relevant figures:* 5.5, 5.14, 5.21

**Table 8.4**  General features of glacial facies

*Deposition:* takes place in a variety of environments: beneath glaciers of various types, in glacial lakes, on glacial outwash plains and glacio-marine shelves and basins, and by a variety of processes including moving and melting glaciers, melt-water streams, melt-water density currents and icebergs.

*Lithologies:* polymictic conglomerates termed tillite, sandstones, muddy sediments with dispersed clasts (dropstones).

*Texture:* poorly-sorted, matrix-supported conglomerates (tillites), angular clasts possibly with striations and facets, and elongate clasts possibly showing preferred orientation.

*Structures:* tillites generally massive but some layering may occur; rhythmically laminated ('varved') muddy sediments common; fluvioglacial sandstones show cross-bedding and lamination, flat bedding, scours and channels.

*Fossils:* generally absent (or derived), except in glaciomarine sediments.

*Geometry:* tillites laterally extensive.

*Facies sequences:* no typical sequence.

*Relevant section:* 4.6.

*Relevant figures:* 4.5, **8.2.**

**Fig. 8.2**  Rhythmically-laminated mudrock containing a boulder (dropstone), dropped by melting iceberg. Ordovician, West Africa.

103

**Table 8.5** General features of deltaic facies

*Deposition:* complex, there are several types of delta (lobate and birdfoot especially), and many deltaic subenvironments (distributary channels and levees, swamps and lakes, mouth and distal bars, interdistributary bays and prodelta slope). Many deltas are river-dominated but reworking and redistribution by marine processes can be important.

*Lithologies:* mainly sandstones (often lithic) through muddy sandstones, sandy mudrocks to mudrocks; also coal seams and ironstones.

*Textures:* not diagnostic, typically average sorting and rounding of sand grains.

*Structures:* cross-bedding of various types in the sandstones, flat bedding and channels common. Finer sediments show flaser and wavy bedding. Some mudrocks contain rootlets; nodules of siderite common.

*Fossils:* marine fossils in some mudrocks and sandstones, others with non-marine fossils, especially bivalves. Plants common.

*Palaeocurrents:* mainly directed offshore but may be shore-parallel or onshore if much marine reworking.

*Geometry:* sand bodies vary from ribbons to sheets depending on delta type.

*Facies sequence:* these typically consist of coarsening upward units (mudrock to sandstone), through delta progradation, capped by a seatearth and coal; there are many variations however, particularly at the top of such units.

*Relevant sections:* 3.2, 3.4, 5.2, 5.3

*Relevant figures:* 5.3 to 5.16, 5.18, 8.3

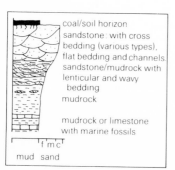

coal/soil horizon
sandstone: with cross bedding (various types), flat bedding and channels.
sandstone/mudrock with lenticular and wavy bedding
mudrock

mudrock or limestone with marine fossils

f m c
mud    sand

**Fig. 8.3** Typical (simple) coarsening-up sequence produced by delta progradation (thicknesses range from 10 to 30 m or more). There can be much variation in such a sequence, particularly towards the top if interdistributary bays are developed.

**Table 8.6**  General features of shallow-marine siliciclastic facies

*Deposition:* takes place in a variety of environments and subenvironments including tidal flat, beach, barrier island, lagoon and nearshore to offshore shelf. Waves, tidal and storm currents are the most important processes.

*Lithologies:* mainly sandstones (often quartz arenites) through muddy sandstone, sandy mudrocks to mudrocks; also thin conglomerates.

*Textures:* not diagnostic although sandstones often have well-rounded and well-sorted grains.

*Structures:* in the sandstones: cross-bedding, possibly herring-bone in character and with reactivation surfaces, flat bedding (in truncated sets if beach), wave-formed and current ripples and cross-lamination, flaser and lenticular bedding, desiccation cracks, thin graded sandstones of storm current origin; mudrocks may contain pyrite nodules; bioturbation and various trace fossils common.

*Fossils:* marine faunas with diversity dependent on salinity, level of turbulence, substrate, etc.

*Palaeocurrents:* variable, parallel to and normal to shoreline, unimodal, bimodal or polymodal.

*Geometry:* linear sand bodies if barrier or beach, sheet sands if extensive epeiric-sea platform.

*Facies sequences:* vary considerably depending on precise environment and sea-level history (rising or falling); both fining up and coarsening up units occur.

*Facies associations:* limestones, ironstones and phosphates may occur within shallow-marine siliciclastic facies.

*Relevant sections:* 3.2, 3.4, 5.3, 5.6

*Relevant figures:* 5.4 to 5.16, 5.19, 5.39 to 5.45, **8.4, 8.5**

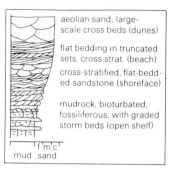

**Fig. 8.4** One example of a shallow-marine siliciclastic sequence: coarsening up sequence resulting from progradation of a beach/barrier shoreline. Typical thickness 10 m or more.

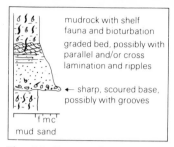

**Fig. 8.5** Features of a storm bed, common in shallow-marine shelf sequences, siliciclastic or carbonate. Storm beds are generally less than 20 cm in thickness.

---

**Table 8.7** General features of deeper-marine siliciclastic facies

*Deposition:* takes place on submarine slopes, submarine fans and in basins of many types, particularly by turbidity currents, debris flows, contour currents and deposition from suspension.

*Lithologies:* sandstones (often greywacke in composition) and mudrocks; also conglomerates.

*Texture:* not diagnostic; sandstones often matrix-rich; conglomerates mostly matrix-supported and of debris flow origin.

*Structures:* in sandstones of turbidity current origin: graded beds (interbedded with hemipelagic mudrocks) which may show 'Bouma' sequence of structures (Fig. 8.6); sole marks common, channels perhaps large-scale, also slump and dewatering structures. Some sandstones may be massive. Mudrocks may be finely laminated.

*Fossils:* mudrocks chiefly contain pelagic fossils; interbedded sandstones may contain derived shallow water fossils.

*Palaeocurrents:* variable, may be downslope or along basin axis.

*Facies sequences:* turbidite successions may show upwards coarsening and thickening of sandstone beds, or upwards fining and thinning.

*Relevant sections:* 3.2, 5.2, 5.3

*Relevant figures:* 5.1, 5.2, 5.31, 5.32, **8.6, 8.7, 8.8**

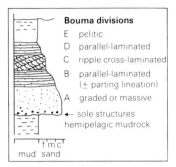

**Bouma divisions**

E  pelitic

D  parallel-laminated

C  ripple cross-laminated

B  parallel-laminated
   (± parting lineation)

A  graded or massive

← sole structures
hemipelagic mudrock

f m c
mud  sand

**Fig. 8.6**  An ideal turbidite, with Bouma divisions. In many turbidites, not all divisions are developed; AE, BCE, CDE and CE sequences are common. Turbidite beds range from a few centimetres to a metre or more in thickness.

**Fig. 8.7**  A turbidite bed showing an ABC sequence: a lower coarse massive division (A), overlain by a finer-grained parallel-laminated division (B), and then the upper part consisting of cross-laminations with a convolution at the top (C division). This turbidite is a bioclastic limestone. Devonian, S.W. England.

**Fig. 8.8 (below)**  Turbidite beds regularly interbedded with hemipelagic mudrock. These turbidites are limestones (full of shallow-water fossils) and the succession is inverted. Devonian, S.W. England.

**Table 8.8**  General features of shallow-marine carbonate facies

*Deposition:* takes place in a variety of environments and subenvironments: tidal flats, beaches, barriers, lagoons, nearshore to offshore shelves and platforms, epeiric shelf seas, submarine shoals and reefs (shelf-margin and patch reefs especially). Biological and biochemical processes are largely responsible for formation and deposition of sediment, although physical processes of waves, tidal and storm currents are important.

*Lithologies:* many types of limestones, especially biosparites, oosparites, biomicrites and pelleted limestones; also dolomites. Limestones may be silicified. Evaporites, especially sulphates (or their replacements) may be associated.

*Textures:* diverse.

*Structures:* diverse including cross-bedding, flat bedding, scours, ripples, desiccation cracks, stromatolites, fenestrae, stromatactis and stylolites; reef limestones: massive and unbedded, many organisms in growth position.

*Fossils:* vary from diverse and abundant in normal marine facies to restricted and rare in hypersaline or hyposaline facies.

*Palaeocurrents:* variable: parallel and normal to shoreline.

*Facies sequences:* many types but shallowing-up sequences are common (Fig. 8.9).

*Relevant sections:* 3.5, 5.3, 5.4, 5.6, 6.1 to 6.4

*Relevant figures:* 3.3 to 3.7, 5.23 to 5.30, 5.39 to 5.45, 6.1 to 6.4, **8.9**

**Table 8.9** General features of deeper-water carbonates and other pelagic facies

---

*Deposition:* takes place in deeper-water epeiric seas, outer shelves and platforms, submarine slopes, in basins of many types and on ridges and banks within basinal areas. Deposition is from suspension and by resedimentation processes.

*Lithologies:* pelagic limestones are usually fine grained with dominantly pelagic fauna; limestone turbidites are coarser grained and consist largely of shallow water fossils; cherts, phosphorites, iron-manganese nodules and enrichments, hemipelagic mudrocks.

*Structures:* pelagic limestones: often nodular, hardgrounds common together with sheet cracks and neptunian dykes, stylolites common; turbidite limestones: graded bedding and other structures (sole and internal) as in Fig. 8.6 although often less well developed; bedded cherts: may be graded and laminated. Pelagic sediments may be slump folded and brecciated.

*Fossils:* pelagic fossils dominate; derived shallow-water fossils in limestone turbidites.

*Facies sequences:* no typical sequences; pelagic facies may overlie or underlie turbidite successions or follow platform carbonates. Pelagic facies may be associated with volcaniclastic sediments and pillow lavas.

*Relevant sections:* 3.5, 3.8

*Relevant figures:* 3.11, 5.31, 5.32, 5.35, 5.36, **8.6 to 8.9**

---

palaeokarstic surface and/or sabkha evaporites, may be replaced (subaerial, supratidal)
fenestral micrites and cryptalgal laminites (tidal flat)
biomicrites, restricted fauna (lagoonal)

oolitic/bioclastic grainstone + cross strat. (shoal, barrier)

biomicrite, rich and varied fauna, with graded storm beds (open shelf)

f m c
mud    sand

**Fig. 8.9** An example of a shallowing-up carbonate sequence; there are many variations on this generalized scheme, particularly if reefs are developed.

**Table 8.10** General features of volcaniclastic facies

*Deposition:* takes place in subaerial and submarine (shallow or deep) environments by pyroclastic fall-out, volcaniclastic flows such as ignimbrites, lahars and base surges, and reworking and resediment-ation by waves, tidal, storm and turbidity currents.

*Lithologies:* tuffs, lapillistones, agglomerates and breccias.

*Textures:* diverse, include welding in ignimbrites and matrix-support fabric in lahar deposits.

*Structures:* include grading in air-fall tuffs, current and wave structures in reworked and redeposited tuffs, planar and cross-bedding in base surge tuffs.

*Fossils:* do occur although may be rare.

*Facies associations:* submarine volcaniclastics frequently associated with pillow lavas, cherts and other pelagic sediments.

*Relevant section:* 3.11

*Relevant figures:* 3.13 to 3.15

*Tables:* 3.4, 3.5

# References and further reading

*Books containing sections on field techniques include:*

BOUMA, A.H. (1962) *Sedimentology of some Flysch Deposits: a Graphic Approach to Facies Interpretation*, p. 168. Elsevier, Amsterdam.

BOUMA, A.H. (1969) *Methods for the Study of Sedimentary Structures*, p. 458. Wiley-Interscience, New York.

CARVER, R.E. (Ed.) (1971) *Procedures in Sedimentary Petrology*, p. 653. Wiley, New York.

COMPTON, R.R. (1962) *Manual of Field Geology*, p. 378. Wiley, New York.

MÜLLER, G. (1967) *Methods in Sedimentary Petrology*, p. 283. Schweizerbart, Stuttgart.

*General textbooks concerned with sedimentary rocks include:*

BLATT, H., MIDDLETON, G.V., MURRAY, R.C. (1980) *Origin of Sedimentary Rocks*, p. 800. Prentice Hall, Englewood Cliffs, N.J.

FOLK, R.L. (1974) *Petrology of Sedimentary Rocks*, p. 159. Hemphill Pub. Co., Austin, Texas.

FRIEDMAN, G.M. & SAUNDERS, J.E. (1978) *Principles of Sedimentology*, p. 792. Wiley, New York.

FÜCHTBAUER, H. (1974) *Sediments and Sedimentary Rocks*, p. 464. Schweizerbart, Stuttgart.

GREENSMITH, J.T. (1978) *Petrology of the Sedimentary Rocks*, p. 241. George Allen & Unwin, London.

PETTIJOHN, F.J. (1975) *Sedimentary Rocks*, p. 628. Harper & Row, New York.

TUCKER, M.E. (1981) *Sedimentary Petrology: an introduction*, p. 252. Blackwells, Oxford.

*General textbooks concerned with depositional environments and facies analysis include:*

HALLAM, A. (1981) *Facies Interpretation and the Stratigraphic Record*, p. 291. Freeman, Oxford.

READING, H.G. (Ed.) (1978) *Sedimentary Environments and Facies*, p. 557. Blackwells, Oxford.

REINECK, H.E. & SINGH, I.B. (1980) *Depositional Sedimentary Environments*, p. 549. Springer-Verlag, Berlin.

RIGBY, J.K. & HAMBLIN, W.K. (Eds.) (1972) *Recognition of Ancient Sedimentary Environments*, p. 340. Spec. Publ. Soc. econ. Palaeont. Miner., *16*, Tulsa.

SELLEY, R.C. (1978) *Ancient Sedimentary Environments*, p. 287. Chapman & Hall, London.

WALKER, R.G. (Ed.) (1979) *Facies Models*, p. 211. Geoscience Canada.

*Books discussing particular sedimentary rock groups include:*

BATHURST, R.G.C. (1975) *Carbonate Sediments and their Diagenesis.* p. 658. Elsevier, Amsterdam.

KIRKLAND, D.W. & EVANS, R. (Eds.) (1973) *Marine Evaporites: Origins, Diagenesis and Geochemistry,* p. 426. Dowden, Hutchinson & Ross, Stroudsburg.

LEPP, H. (Ed.) (1975) *Geochemistry of Iron,* p. 464. Dowden, Hutchinson & Ross, Stroudsburg.

PETTIJOHN, F.J., POTTER, P.E. & SIEVER, R. (1973) *Sand and Sandstone,* p. 617. Springer-Verlag, Berlin.

POTTER, P.E., MAYNARD, J.B. & PRYOR, W.A. (1980) *Sedimentology of Shale,* p. 270. Springer-Verlag, Berlin.

WILSON, J.L. (1975) *Carbonate Facies in Geologic History,* p. 471. Springer-Verlag, Berlin.

*Other pertinent textbooks include:*

CONYBEARE, C.E.B. & CROOK, K.A.W. (1968) *Manual of Sedimentary Structures,* p. 327. Australian Dept. Natl. Development. *Bull. Bur. Min. Res., Geol. & Geophysics, 102.*

PETTIJOHN, F.J. & POTTER, P.E. (1964) *Atlas and Glossary of Primary Sedimentary Structures,* p. 370. Springer-Verlag, Berlin.

POTTER, P.E. & PETTIJOHN, F.J. (1977) *Palaeocurrents and Basin Analysis,* p. 425. Springer-Verlag, Berlin.

RAGAN, D.M. (1973) *Structural Geology* p. 208. Wiley, New York.

FREY, R.W. (Ed.) (1975) *The Study of Trace Fossils,* p. 562. Springer-Verlag, New York.

*For stratigraphic procedure consult:*

HEDBERG, H.D. (Ed.) (1976) *International Stratigraphic Guide,* p. 200. Wiley Interscience.

HOLLAND, C.H. et al. (1978) *A guide to Stratigraphical Procedure. Geol. Soc. Lond. Spec. Rep.* 11, p. 18.